Lecture Notes in Artificial Intelligence   1221

Subseries of Lecture Notes in Computer Science
Edited by J. G. Carbonell and J. Siekmann

Lecture Notes in Computer Science

Edited by G. Goos, J. Hartmanis and J. van Leeuwen

Springer

Berlin
Heidelberg
New York
Barcelona
Budapest
Hong Kong
London
Milan
Paris
Santa Clara
Singapore
Tokyo

Gerhard Weiß (Ed.)

Distributed Artificial Intelligence Meets Machine Learning

Learning in Multi-Agent Environments

ECAI'96 Workshop LDAIS
Budapest, Hungary, August 13, 1996
ICMAS'96 Workshop LIOME
Kyoto, Japan, December 10, 1996
Selected Papers

Springer

Series Editors
Jaime G. Carbonell, Carnegie Mellon University, Pittsburgh, PA, USA
Jörg Siekmann, University of Saarland, Saarbrücken, Germany

Volume Editor

Gerhard Weiß
Institut für Informatik, Technische Universität München
D-80290 München, Germany
E-mail: weissg@informatik.tu-muenchen.de

Cataloging-in-Publication Data applied for

Die Deutsche Bibliothek - CIP-Einheitsaufnahme

Distributed artificial intelligence meets machine learning : learning
in multi-agent environments ; selected papers / ECAI '96, Workshop
LDAIS, Budapest, Hungary, August 13, 1996 ; ICMAS '96,
Workshop LIOME, Kyoto, Japan, December 10, 1996. Gerhard Weiss
(ed.). - Berlin ; Heidelberg ; New York ; Barcelona ; Budapest ; Hong
Kong ; London ; Milan ; Paris ; Santa Clara ; Singapore ; Tokyo :
Springer, 1997
   (Lecture notes in computer science ; Vol. 1221 : Lecture notes in
   artificial intelligence)
   ISBN 3-540-62934-3

CR Subject Classification (1991): I.2, I.6, D.3.2

ISBN 3-540-62934-3 Springer-Verlag Berlin Heidelberg New York

Typesetting: Camera ready by author
SPIN 10549569   06/3142 – 5 4 3 2 1 0   Printed on acid-free paper

Preface

The intersection of distributed artificial intelligence and machine learning constitutes a relatively young but important area of research that has received steadily increasing attention in the past years. The reason for this attention is largely based on the insight that the complexity of the systems studied in distributed artificial intelligence often makes it extremely difficult or even impossible to correctly and completely specify their behavioral repertoires and their dynamics. It is therefore broadly agreed that these systems should be equipped with the ability to learn, that is, to improve their future performance on their own. This book documents current and ongoing developments in the area of learning in distributed artificial intelligence systems.

The book contains selected, revised, and extended versions of sixteen papers that were first presented at two related workshops held at the Twelfth European Conference on Artificial Intelligence (ECAI-96, Budapest, Hungary, August 11–16, 1996) and the Second International Conference on Multiagent Systems (ICMAS-96, Kyoto, Japan, December 9–13, 1996). These were the ECAI-96 workshop on "Learning in Distributed Artificial Intelligence Systems" (LDAIS) and the ICMAS-96 workshop on "Learning, Interaction, and Organization in Multiagent Environments" (LIOME). Additionally, the book contains the invited talk by Munindar Singh and Michael Huhns presented at the LDAIS workshop and a Reader's Guide. Forty-one papers were submitted to these two workshops, from nine different countries: fifteen from Japan, nine from the USA, five from Germany, four from the UK, two each from Spain, Italy, and Switzerland, and one each from Romania and Sweden. Each paper submitted to the LDAIS or the LIOME workshop was reviewed by at least two experts. Twenty-five of the submitted papers were accepted for presentation at the workshops, and sixteen of them were selected for subsequent publication in this book. The purpose of both workshops was to bring together researchers and practitioners with an active interest in both distributed artificial intelligence and machine learning, and to serve as a forum for discussing existing work, exchanging expertise, and developing new ideas and perspectives. Further information on these workshops and their organization is provided on the following pages.

The papers included in this book reflect the broad spectrum of learning in distributed artificial intelligence systems and the progress made in this area. It is my hope that this book will serve as a valuable source of information and inspiration for the reader, and that it will lead to further development and progress in this fascinating area.

Acknowledgements. I would like to thank all the people who contributed to these workshops and thereby made this book possible at all. I am particularly grateful to the committee members of both workshops for their advice and review activities, to the additional reviewers, to Munindar Singh and Michael Huhns for giving an invited talk at the LDAIS workshop, to the speakers for presenting their work, to the workshop attendees for their interest, and to the authors for their willingness to prepare revised and extended versions of their workshop contributions. The LIOME workshop would not have been possible without the commitment and engagement of Hiroshi Ishiguro. Last not least, I am indebt to Alfred Hofmann of Springer-Verlag for his assistance and support during the whole book project.

February 1997   Gerhard Weiß

Organization of the ECAI-96 Workshop on Learning in Distributed Artificial Intelligence Systems (LDAIS)

Organizing Committee

Gerhard Weiß (Chair)	Technische Universität München, Germany
Michael Huhns	University of South Carolina, USA
Toru Ishida	Kyoto University, Japan
Gerhard Kraetzschmar	Universität Ulm, Germany
Victor Lesser	University of Massachusetts, USA
Jeffrey Rosenschein	The Hebrew University, Israel
Gheorghe Tecuci	George Mason University, USA, & Romanian Academy, Romania
Walter Van de Velde	Vrije Universiteit Brussel, Belgium

Additional Reviewers

Ciara Byrne	Trinity College, Dublin, Ireland
Diana Gordon	Navy Center for Applied Research in AI, USA
Thomas Haynes	University of Tulsa, USA
Tuomas Sandholm	University of Massachusetts at Amherst, USA
Ming Tan	GTE Laboratories Inc., USA
Moshe Tennenholtz	Technion - Israel Institute of Technology, Israel

Invited Speakers

Munindar Singh (North Carolina State University, USA) and Michael Huhns (University of South Carolina, USA), "Challenges for Machine Learning in Cooperative Information Systems"

Organization of the ICMAS-96 Workshop on Learning, Interaction and Organization in Multiagent Environments (LIOME)

Organizing Committee

Yukinori Kakazu	Hokkaido University, Japan
Toru Ishida	Kyoto University, Japan
Gerhard Weiß	Technische Universität München, Germany

Program Committee

Minoru Asada	Osaka University, Japan
Yves Demazeau	Laboratoire Leibniz, Institut IMAG, France
Ed Durfee	University of Michigan, USA
Thomas Haynes	University of Tulsa, USA
Michael Huhns	University of South Carolina, USA
Hiroshi Ishiguro	Kyoto University, Japan
Victor Lesser	University of Massachusetts, USA
Jeffrey Rosenschein	The Hebrew University, Israel
Sandip Sen	University of Tulsa, USA
Keiji Suzuki	Hokkaido University, Japan
Tatsuo Unemi	Soka University, Japan

Contents

Part III: Learning, Communication and Understanding

Reader's Guide

Gerhard Weiß

Institut für Informatik, Technische Universität München
D-80290 München, Germany
weissg@informatik.tu-muenchen.de

Learning in multiagent environments establishes a relatively young research and application area that has achieved steadily increasing attention in the past years. The importance of this area is broadly acknowledged in the distributed artificial intelligence community as well as the machine learning community. The seventeen papers included in this book reflect current major developments in this area. In the following, these papers are briefly characterized and motivated.

In the first paper, which is included in this introductory part, *Munindar Singh and Michael Huhns* provide a broad overview and careful discussion of the central *challenges for machine learning in cooperative information systems*. Cooperative information systems are multiagent systems that work effectively within modern – distributed, large, open, and heterogeneous – information environments. These systems constitute an increasingly important and popular research area at the intersection of distributed artificial intelligence, distributed databases, and distributed computing (see, e.g., (CoopIS, 1996)). The need for learning in these multiagent systems results from their enormous complexity and the difficulty in equipping them with a robust but flexible structure and functionality. Singh and Huhns distinguish three major machine learning challenges: extraction of semantics, coordination and collaboration, and abstractions and structure of cooperative information systems. The paper is rounded off by an overview of different categories of learning and their potential applications to problems in cooperative information systems.

The remaining sixteen papers are grouped according to their main focus into three parts:

- Learning, cooperation and competition. This part includes work whose main focus is on the question how multiple agents can learn cooperative and/or competitive behavior (e.g., in order to optimally share resources or to maximize one own's profit). This work is much concerned with the development and adaptation of data-flow and control patterns that improve the interactions among several agents.
- Learning about/from other agents and the world. Here the main focus is on the question how learning conducted by an agent can be influenced (e.g., initiated, accelerated, re-directed, made possible at all, and so forth) by other agents. This work is much concerned with the prediction of the behavior of

other agents (including their preferences, strategies, intentions, goals, etc), with the improvement and refinement of an agent's behavior by interacting with and observing other agents, and with the development of a common view of the world.

- Learning, communication and understanding. Here the main focus is on the question how learning on the one hand and communication and mutual understanding on the other hand are related to each other. This work is much concerned with the requirements on the agents' ability to effectively exchange useful information and to develop a shared meaning of the information exchanged.

(It is important to see that these parts are not orthogonal, but complement one another. For instance, agents may learn to cooperate by learning about each other's abilities, and in order to learn from one another the agents typically have to communicate with each other.)

Part I: Learning, Cooperation and Competition

Ono and Fukumoto – Modular multiagent learning. A basic problem arising in cooperative and competitive learning tasks is the complexity of learning. The more agents are involved in such a learning task, the faster grows each individual agent's learning space. The reason for this is that the number of potential agent–agent combinations that have to be taken into consideration by an agent in order to find out appropriate cooperation or competition patterns grows exponentially in the number of agents. This problem has been attacked by Norihiko Ono and Kenji Fukumoto. Their multiagent reinforcement learning approach is based on a variant of the modular Q-learning architecture introduced by Whitehead et al. (1993). The main idea underlying this approach is that each agent's learning component is decomposed into independent modules, each focusing on a separate aspect of the environment or the task to be solved. The learning results of these components are combined by a mediator module using a simple heuristic procedure. As noted by the authors, a major advantage of their approach compared to others like those described in e.g. (Drogoul et al., 1991; Tan, 1993) is that it scales better to more difficult multiagent learning problems.

Versino and Gambardella – Learning real team solutions. The key problem that every learning system has to solve is the so-called credit-assignment problem (Minsky, 1961), that is, the problem of properly assigning a received learning feedback – credit or blame – for performance changes to the contributing system activities. Although this problem has been traditionally studied in the context of single-agent systems, it is also present in multiagent contexts (Weiß, 1996). What makes the solution of this problem particularly difficult in multiagent contexts is that often only a single learning feedback is available for a set of activities carried out by different agents pursuing a shared learning goal. An important question therefore is how several agents can learn to what extent each of their activities contributed to a learning feedback that reflects the quality of the agents' overall team behavior, but not the quality of each individual agent's behavior. This question is addressed in the paper of Cristina Versino and Luca Maria Gambardella in the the context of self-programming robots. In what they are particularly interested in is the learning of "real" team solutions, that is, solutions whose shape strongly depends on team properties like the team size or the team composition, and they present and investigate a multiagent reinforcement learning algorithm which aims at the optimization of team performance instead of individual performance.

Davidsson – Learning by anticipation. Rosen (1985) introduced the concept of anticipatory systems which use high-level (symbolic) world models in order to predict future states and to guide low-level (reactive) behavior. A system learns by developing and refining a world model and by modifying its behavioral preferences according to this model. Based on this concept, Paul Davidsson presents an agent architecture called ALQAAA ("A Linearly Quasi-Anticipatory Agent Architecture") and investigates its use in competitive and cooperative multiagent settings. Here the agents are assumed to learn independent of each other, although they may take each others' activities into consideration by integrating them into their world models. The independence among the agents also implies that there is no explicit communication among the agents which influences their individual learning processes.

Ono and Fukuta – Multiagent learning in continuous environments. Most of the work available on multiagent learning is based on the assumption that the agents act in discrete environments. This is an idealistic assumption from the point of view of real-world interaction, at least if no mechanism for the transformation of continuous sensory input into a discrete representation is provided together with the discrete learning algorithm under consideration. Starting out from the observation that in general it is not clear whether learning algorithms designed for discrete environments can be successfully applied in continuous environments, Norihiko Ono and Yoshihiro Fukuta describe an approach to multiagent reinforcement learning which is intended to overcome this limitation. This approach is based on a CMAC-based Q-learning method proposed by Lin (1992). (CMAC, which stands for "Cerebellar Model Articulation Computer" (Albus, 1981), is a biologically motivated model of the cerebellum which is widely used in the area of function approximation.)

Schmidhuber and Zhao – Lifelong learning in multiagent settings. Jürgen Schmidhuber and Jieyu Zhao describe a general learning principle called "realistic reinforcement learning" or "lifelong learning" and show how it can be applied in multiagent environments. According to this principle it is assumed that an agent executes a lifelong action sequence in a complex environment and tries to optimize the cumulative learning feedback at each time during its life. A basic idea underlying this principle is that learning algorithms themselves are considered as actions. This means that the learning algorithms applied by the agents are evaluated in exactly the same way than any other actions, and that there is no longer a distinction between learning, meta-learning, meta-meta-learning, and so forth. A property of this principle which is of particular interest from the point of view of multiagent learning is that cooperation among the agents is likely to be inherently enforced if all agents try to improve the same learning feedback and none of them is able to improve this feedback by itself.

Ye and Tsotsos – Learning to search for objects. Yiming Ye and John Tsotsos formulate the multiagent object search task and discuss various issues of learning arising in this task domain. According to this formulation, several agents have to cooperate in searching for a 3D object in an unknown 3D environment. The authors prove that the object search task is NP-hard in both multiagent and single-agent settings. Several challenging issues like communication and planning are identified which reveal several demands on the learning abilities of the individual agents. The work of Ye and Tsotsos is loosely related to the various approaches on multiagent and distributed search like, e.g., (Ishida & Korf, 1991; Pollack and Ringuette, 1990).

Bazzan – Evolution of coordination strategies. Ana Bazzan concentrates on the question how several agents can achieve both a stable and a favorable coordination of their activities without requiring extensive communication. She proposes a learning approach which employs the evolutionary mechanisms of mutation and selection in order to find coordination strategies that are stable in a game-theoretic sense. According to this evolutionary perspective, the individual agents may be considered as pursuing one common learning goal, although they do not explicitly interact by communicating with each other (the information exchange is reduced to the application of the evolutionary operators). The work of Bazzan is related to several other appoaches to learning in multiagent systems that follow, or at least are inspired by, the principle biological evolution; for instance, see (Bull & Forgarty, 1996; Grefenstette & Daley, 1996; Haynes & Sen, 1996) and also (Schmidhuber and Zhao, in this volume).

Part II: Learning About/From Other Agents and the World

Nadella and Sen – Learning to play soccer. Most approaches to multiagent learning deal with learning to cooperate or with learning to compete, but not with both. In real-world multiagent domains, however, often both cooperative and competitive agents are present, and the individual agents therefore should be capable of both types of learning. In other words, in such domains agents have to learn to be in successful cooperation in order to achieve shared or complementary goals, and at the same time they have to learn to be in successful competition against other agents which pursue other goals being in contradiction to their own ones. This observation was the starting point for the multiagent learning approach described by Rajani Nadella and Sandip Sen. As an application domain they have chosen the robotic soccer domain because here both cooperative learning (within a team) and competitive learning (between teams) can be well studied. The authors developed a soccer simulator which allows them to evaluate cooperative and competitive strategies learnt by the robots. A basic characteristic of this approach is that cooperation and competititon is achieved by learning from other agents and about oneself. (This is the reason why this contribution has been included in part II, although it would have been also a good candidate for part I of this volume.) The work of Nadella and Sen was partially inspired by the robot world cup initiate; see (Kitano et al., 1995). A related work on learning in a robotic soccer domain was presented by Stone and Veloso (1996) who also distinguish between learning to cooperate (cooperative learning) and learning to compete (adversarial learning).

Dragoni and Giorgini – Distributed elicitation of knowledge. How can several agents develop a uniform view of the world, if they have different knowledge and inference abilities? Starting out from this question, Aldo Franco Dragoni and Paolo Giorgini developed a model of distributed belief revision based on Bayesian conditioning that enables agents to cope with inconsistent and incomplete information. The model explicitly takes into consideration that different agents may have different and conflicting points of view of their world and of each other. The basic idea underlying this model is to endow the agents with the ability to learn the other agents' and one own's relative reliabilities in the course of mutual interaction. Experimental results are presented that illustrate how distributed belief revision may influence the reliability of the individual agents and the quality of the individual agents' information bases. The agents may be considered as having one single learning goal – namely, to develop a more or less uniform view of the world –, where interaction among the agents is necessary for achieving this goal. Related works on distributed belief revision can be found in, e.g., (Galliers, 1991; Gaspar, 1991; Jennings et al., 1994).

Terabe et al. – Organizational learning. "Organizational learning" is a term which originally steems from economics. Here organizational learning constitutes a traditional and well established subject of study, and refers to the learning processes conducted by and in an organization like a business company or a state institution that increase its competitiveness, productivity, and innovativeness in uncertain markets and changing technological environments. (A bibliography on organizational learning is included in (Weiß, 1996).) Masahiro Terabe, Takashi Washio, Osamu Katai and Tetsuo Sawaragi offer an approach to organizational learning from the perspective of distributed artificial intelligence. According to their approach, two types of knowledge that each agent possesses are distinguished: knowledge about the task to be solved by the agents ("task knowledge"), and knowledge about the agents and their abilities ("organizational knowledge"). The agents learn about other agents and, hence, improve their organzational knowledge, in order to improve their interaction patterns and to be able to generate more efficient task solutions. This work would be an equally well candidate for the part I of this volume, because organizational learning always has to do with multiagent cooperation. Related work on organizational learning is presented in e.g. (Nagendra Prasad et al., 1995; Weiß, 1994); the reader is also refered to (Carley & Prietula, 1994) where the issues of computational organization are treated from a broader perspective.

Plaza et al. – Multiagent Case-Based Learning. Enric Plaza, Josep Lluis Arcos

6

and Francisco Martin propose two multiagent cooperation modes based on case-based reasoning and learning. According to the distributed mode, an agents sends a problem to be solved to another agent which then applies its own case-based reasoning and learning method and its own case base in order to generate a solution. According to the collective mode, an agent sends a problem plus its case-based reasoning method to another agent which then tries to solve the problem by applying the sending agent's method plus its own case base. In both modes the agents share the experience gained by them in the course of learning, where the learning process itself occurs in a decentralized way. Case-based reasoning and learning are techniques for the representation, maintainance and reuse of knowledge which has been intensively studied in machine learning since many years. Further case-based learning approaches in multiagent contexts are described in, e.g., (Haynes et al., 1996; Nagendra Prasad et al., 1996) and (Ohko et al., in this volume).

Lenzmann and Wachsmuth – Multiagent user adaptation. Intelligent user interfaces have achieved increasing attention in the past decade (see, e.g., Sullivan & Tyler, 1990). Britta Lenzmann and Ipke Wachsmuth present a multiagent perspective of an adaptive user interface. A main idea underlying this work is that users of computers and software programs have different and time-varying preferences, and that a promising way of making a working environment more comfortable is to provide an interface that automatically identifies the preferences of the individual users. The proposed interface consists of several agents, each corresponding to a possible preference. The agents interact according to a modified variant of the contract-net protocol (Smith, 1980) in order to find out the best suited preference profil for a user. The learning process occuring within such an interface is distributed in the sense that different agents may be responsible for different functions (e.g., there may be an agent responsible for speech and another agent responsible for gesture). Because a user corresponds to an agent who is part of the system environment, the interface can be interpreted as a multiagent system that learns about another agent. This also establishes an important difference to many other works on intelligent interfaces like the one presented by Maes and Kozierok (1993), because here a user interface corresponds to a single-agent system and not to a team of cooperating agents.

Part III: Learning, Communication and Understanding

Davies and Edwards – Communication of inductive inferences. A characteristic of inductive inference is that it is logically unsound, which means that sentences derived by inductive inference may be false (whereas sentences derived by deductive inference are always true). This leads to the question how several agents that perform inductive inference on the basis of their local views of the world could communicate induced hypotheses without generating a logically unsound distributed knowledge base. As a solution to this problem, Winton Davies and Peter Edwards propose to not only communicate the hypotheses, but the hypotheses together with their bounds. Their approach is based on the version space model which was introduced by Mitchell (1978) as a tool for interpreting and analyzing learning as a search problem, and aims at enabling an agent to modify a received induced hypothesis without destroying its correct parts. The authors also discuss implications of their approach to KQML, a popular language and protocol for exchanging information and knowledge (e.g., (Finin et al., 1993)). This agent communication facility only permits the communcation of deductive inferences.

Ohko et al. – Learning and communication load reduction. Agents need to communicate in order to coordinate their activities. The communication load required for coordination is a significant factor in the design of efficient multiagent systems. Takuya Ohko, Kazuo Hiraki and Yuichiro Anzai present a system called LEMMING for task negotiation in multi-robot environments in which the individual robots learn to reduce the amount of communication. This system uses the contract net protocol (Smith, 1980) according to which task announcements are broadcasted to all agents which then respond to the announcements. The basic idea underlying LEMMING is that the individual robots learn to avoid broadcasting whenever possible, and to send messages only to those robots that are expected to be appropriate for task execution. In order to achieve this focused addressing, each robot learns from its experience "cases" (essentially composed of the description of previous task announcements and their solutions) that provide information about the task execution qualities of other agents. Experimental results are presented for a simulated robot environment which illustrate the benefits of case-based learning approach over the standard broadcast approach. The work of Plaza et al. in this volume is another example of an application of case-based learning in multiagent environments.

Friedrich et al. – Multiagent learning and understanding. Holger Friedrich, Michael Kaiser, Oliver Rogella and Rüdiger Dillmann focus on the relationships between learning and mutual understanding in multiagent systems. They identify and illustrate several multiagent learning tasks together with their requirements on inter-agent communication. In particular, they stress the importance of a common ontology in order to make a useful information exchange and learning in multiagent systems possible at all. It is argued that communication and learning is not possible if agents assign different meanings to the same symbols, and therefore the development of a common and shared meaning of symbols is considered as an essential learning task in multiagent contexts. This "shared meaning problem" is closely related to (or might be even considered as the "multiagent or distributed artificial intelligence variant" of) the symbol grounding problem (Harnad, 1990), that is, the problem of grounding the meaning of symbols in the real world. According to the physical grounding hypothesis (Brooks, 1990), which has received particular attention especially in behavior-based artificial intelligence and robotics, the grounding of symbols in the physical world is a necessary condition for building a system that is intelligent. This hypothesis was formulated as a counterpart to the symbol system hypothesis (Newell & Simon, 1976) upon which classical knowledgde-oriented artificial intelligence is based and which states that the ability to handle, manipulate and operate on symbols is a necessary and sufficient condition for general intelligence, independent of the symbols' grounding.

Lacey et al. – Learning and truth. Nicholas Lacey, Keiichi Nakata and Mark Lee approach the topic of learning in multiagent systems from an epistemological perspective. Their contribution does not explicitly focus on communication and understandig, but is very closely related to it. As the reader will notice, there are also tight relationships between this and the contributions of Friedrich et al. and Davies and Edwards in this volume. Agents communicate by exchanging information – knowledge and beliefs – that reflects their partial views of the world (including other agents and themselves). Closely related to the concepts of knowledge and belief is the concept of truth. As it is known from the real life, truth is of relative nature, and an agent may consider a piece of information as being true while at the same time another agent considers it as being false. In epistemology three main theories of truth are identified – the correspondence theory, the coherence theory, and the foundationalism theory –, and an important question is what effects the use of different theories of truth may have in the context of learning multiagent systems. Lacey and his colleagues illustrate how agents having different epistemological backgrounds may draw different conclusions and may learn and acquire different viewpoints, even if they initially have the same information. The resulting variety in the individual agents' information bases is considered as advantageous because it increases the chance of finding better solutions. The authors therefore argue that epistemological assumptions should be explicitly taken into account when designing a multiagent system.

References

Albus, J.S. (1981). Brain, behavior, and robotics. *Byte Book*, Subsidiary of McGraw-Hill, Chapter 6, pp. 139–179.

Brooks, R.A. (1990). Elephants don't play chess. *Robotics and Autonomous Systems*, 6, 3–15.

CoopIS (1996). *Proceedings if the First IFCIS International Conference on Co-operative Information Systems*, IEEE-CS Press.

Bull, L., & Fogarty, T. (1996). Evolutionary computing in cooperative multi-agent environments. In (Sen, 1996, pp. 22–27).

Carley, K.M., & Prietula, M.J. (Eds.) (1994). *Computational organization theory.* Lawrence Erlbaum Associates.

Drogoul, A., Ferber, J., Corbara, B., & Fresneau, D. (1991). A behavioral simulation model for the study of emergent social structures. In *Proceedings of the First European Conference on Artificial Life*.

Finin, T., McKay, D., Fritzson, R., & McEntire, R. (1993). KQML: An information and knowledge exchange protocol. In *International Conference on Building and Sharing of Very Large-Scale Knowledge Bases*.

Galliers, J.R. (1991). Modeling autonomous belief revision in dialogue. In Y. Demazeau & J.-P. Müller (Eds.), *Decentralized Artificial Intelligence 2*. Elsevier.

Gaspar, G. (1991). Communication and belief changes in a society of agents: Towards a formal model of an autonomous agent. In Y. Demazeau & J.-P. Müller (Eds.), *Decentralized Artificial Intelligence 2*. Elsevier.

Grefenstette, J., & Daley, R. (1991). Methods for competitive and cooperative co-evolution. In (Sen, 1996, pp. 45–50).

Harnad, S. (1990). The symbol grounding problem. In *Physica D*, 42, 335–346.

Haynes, T., Lau, K., & Sen, S. (1996). Learning cases to compliment rules for conflict resolution in multiagent systems. In (Sen, 1996, pp. 51–56).

Haynes, T., & Sen, S. (1996). Evolving behavioral strategies in predators and prey. In (Weiß & Sen, 1996, pp. 113–126).

Ishida, T., & Korf, R.E. (1991). Moving target search. In *Proceedings of the Twelfth International Joint Conference on Artificial Intelligence* (pp. 204–210).

Lin, L.-J. (1992). Self-improving reactive agents based on reinforcement learning, planning and teaching. *Machine Learning*, 8, pp. 293f.

Jennings, N.R., Malheiro B., & Oliveira, E. (1994). Belief revision in multi-agent systems. In *Proceedings of the 11th European Conference on Artificial Intelligence* (pp. 294–298).

Kitano, H., Asada, M., Kuniyoshi, Y., Noda, I., & Osawa, E. (1995). Robocup: The robot world cup initiative. *Working Notes on the IJCAI-95 Workshop on Entertainment and AI/Alife* (pp. 19–24).

Maes, P., & Kozierok, R. (1990). Learning interface agents. In *Proceedings of the Eleventh National Conference on Artificial Intelligence* (pp. 459–465).

Minsky, M. (1961). Steps towards artificial intelligence. In *Proceedings of the IRE* (pp. 8–30). Reprinted in E.A. Feigenbaum & J. Feldman (Eds.) (1963), *Computers and thought* (pp. 406–450), McGraw-Hill.

Mitchell, T.M. (1978). *Version spaces: An approach to concept learning*. Ph.D. Thesis. Computer Science Department, Stanford University.

Nagendra Prasad, M.V., Lesser, V.R., & Lander, S. (1995). Learning organizational roles in a heterogeneous multi-agent system. In *Proceedings of the Second International Conference on Multiagent Systems* (pp. 291–298).

Nagendra Prasad, M.V., Lesser, V.R., & Lander, S. (1996). On reasoning and retrieval in distributed case bases. *Journal of Visual and Image Representation* (Special Issue on Digital Libraries), 7(1), 74–87.

Newell, A., & Simon, H.A. (1976). Computer science as empirical inquiry: Symbols and search. *Communications of the ACM*, 19(3), 113–126.

Pollack, M.E., & Ringuette, M. (1990). Introducing the tileworld: experimentally evaluating agent architectures. In *Proceedings of the National Conference on Artificial Intelligence* (pp. 183–189).

Rosen, R. (1985). *Anticipatory systems – Philosophical, mathematical and methodological foundations*. Pergamon Press.

Sen, S. (Ed.) (1996). *Adaptation, coevolution and learning in multiagent systems. Papers from the 1996 AAAI Symposium*. Technical Report SS-96-01. AAAI Press.

Smith, R.G. (1980). The contract net protocol: High-level communication and control in a distributed problem solver. *IEEE Transactions on Computers*, C-29(12), 357–366.

Stone, P., & Veloso, M. (1996). Towards collaborative and adversarial learning: A case study in robotic soccer. In (Sen, 1996, pp. 88–92).

Sullivan, J.W., & Tylor, S.W. (Eds.) (1990). *Intelligent user interfaces.* ACM Press.

Tan, M. (1993). Multi-agent reinforcement learning: Independent vs. cooperative agents. In *Proceedings of the Tenth International Conference on Machine Learning* (pp. 330–337).

Weiß, G. (1994). *Some studies in distributed machine learning and organizational design.* Technical Report FKI-189-94. Institut für Informatik, Technische Universität München.

Weiß, G. (1996). Adaptation and learning in multi-agent systems: Some remarks and a bibliography. In (Weiß & Sen, 1996, pp. 1–21).

Weiß, G., & Sen, S. (Eds.) (1996). *Adaption and learning in multi-agent systems.* Lecture Notes in Artificial Intelligence, Vol. 1042. Springer-Verlag.

Whitehead, S., et al. (1993). Learning multiple goal behavior via task decomposition and dynamic policy merging. In J. H. Connell et al. (Eds.), *Robot learning.* Academic Press.

Challenges for Machine Learning
in
Cooperative Information Systems

Munindar P. Singh[1] * and Michael N. Huhns[2]

[1] Department of Computer Science
North Carolina State University
Raleigh, NC 27695-8206, USA
singh@ncsu.edu
[2] Department of Electrical & Computer Engineering
University of South Carolina
Columbia, SC 29208, USA
huhns@sc.edu

Abstract. *Cooperative Information Systems (CISs)* are multiagent systems with organizational and database abstractions geared to the large **open** heterogeneous information environments of today. CIS is also the name of the associated research area, which has emerged from the synthesis of distributed databases and distributed artificial intelligence. In CIS, software agents mitigate an information environment's heterogeneity by interacting through common protocols, and manage its large size by making intelligent local decisions without centralized control. In order to cope with the dynamism presented by open environments, CIS agents must have the ability to adapt and learn. We discuss some of the most important problems involving learning and adaptivity in CISs, including requirements for reconciling semantics and improving coordination. We present a "customers' view" of learning technology as might find ready application in CISs.

1   Introduction

Due to the proliferation of networking, the desires of almost everyone to be interconnected, and the needs to make data accessible at any time and any place, modern information environments have become large, open, and heterogeneous. They are composed of distributed, largely autonomous, often legacy-based components. *Cooperative Information Systems* introduce software agents into such environments to deal with these characteristics. The agents represent the components in interactions, where they mediate differences and provide a syntactically uniform and semantically consistent middleware. Their greatest difficulty

---

* Munindar P. Singh was partially supported by the NCSU College of Engineering, by the National Science Foundation under grants IRI-9529179 and IRI-9624425, and IBM Corporation.

in achieving uniformity and consistency is the dynamism that open environments introduce.

Open environments are becoming an increasing part of the modern milieu through applications such as information search, electronic commerce, and virtual enterprises. They typically have the following key distinguishing characteristics:

- span enterprise boundaries;
- have components that are heterogeneous in a number of ways, such as the underlying database management systems used, and the semantics associated with the information stored or manipulated;
- comprise information resources that can be added or removed in a loosely structured manner;
- lack global control of the content of those resources, or how that content may be updated; and
- incorporate intricate interdependencies among their components.

To build systems that work effectively within open environments requires balancing their ease of construction and robustness with their flexibility. There are a number of technical difficulties specific to building systems for open environments. Foremost among these are the need to handle the unpredictability in the environment as new components appear, and old ones disappear or change. Since the information components in the environment cannot easily be altered, the agents that represent them must be able to learn and adapt. This provides new challenges for machine learning, as summarized in Table 1:

Traditional Machine Learning	CIS Machine Learning
Agent learns about its environment, which is passive and has no intentions	Agent learns about its environment, which is *active*, because it includes other agents who have intentions, commitments, beliefs, and abilities, and can learn
Agent might have imprecise sensors that cause it to learn inaccurate information about the environment	Agent might deliberately be misled about the environment by other agents

Table 1. Machine Learning for Cooperative Information Systems

Section 2 introduces cooperative information systems, and their quintessential applications. Sections 3, 4, and 5 discuss the machine learning challenges in extracting semantics from passive components, coordinating active components, and abstracting and structuring CISs, respectively. Section 6 concludes with a discussion of the main themes of CIS, and how they relate to machine learning techniques.

2 An Overview of CIS

Cooperative Information Systems (CISs) are an increasingly popular approach that seeks to maximize the above properties through the use of combinations of techniques from distributed artificial intelligence, databases, and distributed computing. The term *cooperative information systems* also refers to the research area that focuses on building such systems.

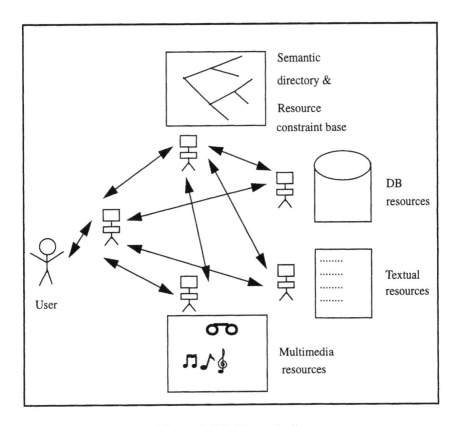

Fig. 1. A CIS Schematically

We define an *agent* as an active, persistent computational entity that can perceive, reason about, and act in its environment, and can communicate with other agents. Agents are autonomous to varying degrees to reflect the autonomy of the information resources or humans whom they represent. Figure 1 shows a CIS schematically. In this figure, we consider an environment consisting of a variety of information resources, coupled with some kind of a semantic directory. The semantic directory contains information about the resources, including any constraints that apply to their joint behavior.

Each component of the environment, as well as the human user(s), is modeled as associated with an agent. The agents capture and enforce the requirements of their associated parties. They interact with one another appropriately, and help achieve the robustness and flexibility in behavior that is required. The charm of agents is that they provide a natural means for acquiring, managing, advertising, finding, fusing, and using information over uncontrollable environments. Further, agents are inherently modular, and can be constructed locally for each resource, provided they satisfy some high-level protocol of interaction.

The applications of CISs are varied. They involve the purely informational ones, such as database access, information malls, workflow management, electronic commerce, and virtual enterprises. They also include physical ones, such as sensor arrays, manufacturing, transportation, energy distribution, and telecommunications.

The above motivates the interest in cooperative information systems. But, as remarked above, CISs involve combining ideas not only from the study of agents, but also from databases and distributed computing. We discuss the specific challenges posed by CISs next. In doing so, we review two of the quintessential applications of CIS: information access and workflow management.

3 Extraction of Semantics

Learning about passive—and often preexisting—components, such as databases and knowledge bases.

Information access involves finding, retrieving, and fusing information from a number of heterogeneous sources. At the level of abstraction that concerns CIS, we are not concerned with network connectivity or the formatting variations of data access languages. Rather, our concern is with the meaning of the information stored. It is possible, and indeed common, that when different databases store information on related topics, each provides a different model of it. The databases might use different terms, e.g., *employee* or *staff*, to refer to the same concept. Worse still, they might use the same term to have different meanings. For example, one database may use *employee* to mean anyone currently on the payroll, whereas another may use *employee* to mean anyone currently receiving benefits. The former will include assigned contractors; the latter will include retirees. Consequently, merging information meaningfully is nontrivial. The problem is exacerbated by advances in communications infrastructure and competitive pressures, because different companies or divisions of a large company, which previously proceeded independently of one another, are now expected to have some linkage with each other.

The linkages can be thought of as semantic mappings between the application (which consumes or produces information), and the various databases. If the application somehow knows that *employee* from a database has one meaning, it can insert appropriate tests to eliminate the records it does not need. Clearly, this approach would be a nightmare to maintain. The slightest changes in a database would require modifying all the applications that access its contents! This would be a fundamental step backward from the very idea of the database

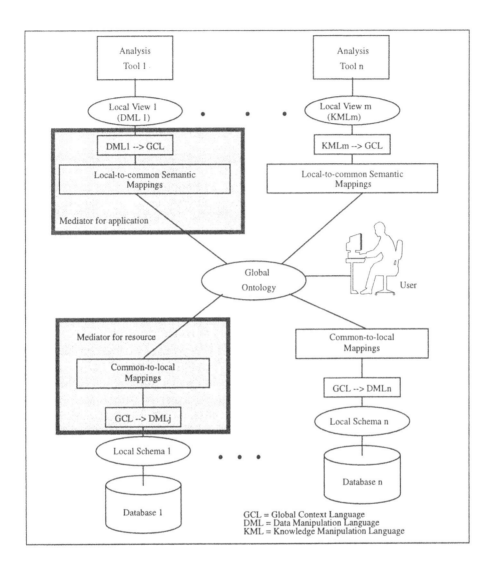

Fig. 2. Mediators

architecture [Elmasri & Navathe, 1994, ch. 1], which sought to separate and shield applications from the storage of data.

A promising approach is to use *mediators* [Wiederhold, 1992]. A mediator is a simplified agent that acts on behalf of a set of information resources or applications. Figure 2 shows a mediator architecture. The basic idea is that the mediator is responsible for mapping the resources or applications to the rest of the world. Mediators thus shield the different components of the system from each other. To construct mediators effectively requires some common representation of the meanings of the resources and applications they connect. Such a knowledge representation is called an *ontology* [Neches *et al.*, 1991]. The main learning challenges associated with ontologies include relationship and concept acquisition. Since these are related to traditional machine learning problems, some progress has already been made on them, but they are far from solved from the perspective of CIS applications.

3.1 Relationship Acquisition

The major problem with ontology-based approaches is the effort required to build them, and to relate different resources and applications to them. In order to extend their world model, agents need to be able to acquire and integrate ontologies autonomously. Agents also should learn the ontologies of other agents. In other cases, tools that assist a human designer are needed. These tools must have a strong machine learning component, to be able to not only relate concepts across databases, but also help identify relationships within an ontology. Such relationships, e.g., generalization or containment, are necessary for CIS query processing approaches, e.g., [Arens *et al.*, 1993; Huhns *et al.*, 1994]. For example, the concept *port* is a generalization of *airport* and can be used to answer queries about airports only if additional restrictions are added.

Further, different domains often have a rich variety of relationships that compose elegantly with each other [Huhns & Stephens, 1989]. To give a simple, albeit somewhat contrived, example, if a person owns a car, and the car contains a wheel, then the person also owns the wheel. These relationships form part of the common sense knowledge that is essential in relating information from different databases: two tables *car-ownership* and *car-parts* in one database may correspond to a single table *auto-part-ownership* in another database.

3.2 Concept Acquisition

We assumed in the above that the concepts that a given database is about are known. A more basic challenge is to identify those concepts. This is potentially useful, but extremely difficult, when dealing with a previously unknown source. It remains useful, and becomes more tractable, when the structure of the database is known, but the structure does not faithfully reflect the meaning of the content. A problem that arises in legacy databases is that they are often misused! For example, databases in the telecommunications industry store information about signal channels. When fiber optic technology was introduced, the databases were not redesigned to capture the new kind of channel. Rather, the existing fields in the databases were overloaded. Consequently, the *conductivity* field may reflect either the conductivity for a copper channel, or the bandwidth for a fiber channel! To access these databases systematically requires knowing what concepts they store, but the concepts are hidden inside the data values. A challenge is to discover the rules for partitioning the mixed up concepts into the correct categories. Some progress is already being made, mostly under the rubric of *data mining*, e.g., [Fayyad, 1996; Shen & Leng, 1996;

Zhang *et al.*, 1996]. An issue that has not drawn much attention is collaborative learning of the concepts. This can be important, because different uses of the data might treat the implicit concepts differently.

4 Coordination and Collaboration

Learning about active components, such as workflows and agents, and their interactions.

4.1 Workflow Acquisition

CISs not only involve retrieving information, but also updating it. Updates are qualitatively more complex than retrievals, because they can potentially introduce inconsistencies. This is especially the case when several databases are involved, and there are subtle interdependencies among them. A *workflow* is a composite activity that accesses different resources and has human interaction to solve some business need.

Traditional databases support so-called ACID transactions, which are computations that are atomic, consistency-preserving, isolated, and durable [Gray & Reuter, 1993]. In other words, a transaction happens entirely or none at all, does not violate consistency, does not expose any partial results, and if successful has permanent results. Transactions are effective in homogeneous and centralized databases, but do not apply in distributed and heterogeneous settings. This is because to ensure the ACID properties requires the component databases to expose their internal control states, and requires locking data items on a database even when those are not in use any more.

This has led to a number of extended transaction models [Bukhres & Elmagarmid, 1996; Elmagarmid, 1992]. Transaction models capture some of the aspects of workflows. Figure 3 gives a trip-planning workflow in the notation of [Buchmann *et al.*, 1992]. This workflow has a number of separate activities, such as opening an account, reserving a flight, booking a hotel, renting a car, and generating a bill. These execute on databases belonging to autonomous organizations, such as different airlines or hotels. Since the airlines make reservations independently of each other and of hotel bookings, the travel agency has to provide the control to make sure air tickets are not bought unnecessarily. Typically, a human would carry out the steps described in Figure 3. Approaches such as [Buchmann *et al.*, 1992] provide a way of representing the dependencies among the steps, and executing them appropriately. However, a major challenge is in determining the structure of workflows, possibly by observing how humans carry them out.

Because there are a large variety of extended transaction models, some so-called "RISC" approaches have been proposed that provide a small set of primitives with which to encode the behavior of different transaction models, e.g.,

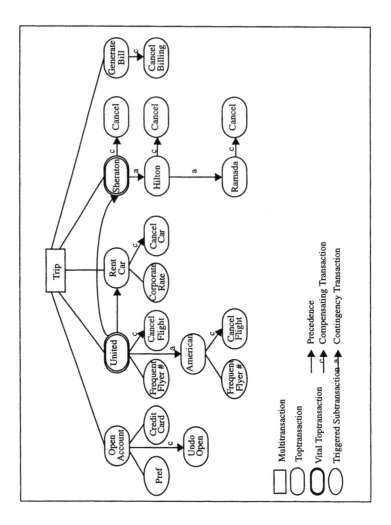

Fig. 3. Workflow for Trip Planning

[Attie *et al.*, 1993; Chrysanthis & Ramamritham, 1994; Singh, 1996]. These approaches provide some variant of a temporal language in which the coordination requirements of the transaction models can be expressed. The approaches of [Attie *et al.*, 1993; Singh, 1996] automatically produce schedules from those specifications. Our challenge can then be framed in terms of how the formal specifications are produced. We believe that the RISC approaches will facilitate learning, because they are declarative and offer a small set of primitives.

4.2   Collaboration Acquisition

Because they are large scale and open, CISs typically involve more than one workflow. However, since these workflows execute on the same resources, they

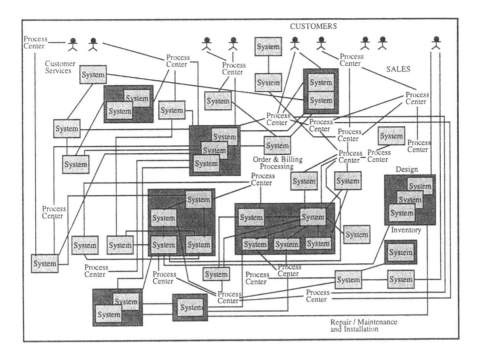

Fig. 4. The Workflow Coordination Required

have a number of interactions. Some of these interactions can be pernicious in that one workflow may cause the failure of another workflow. Some of the interactions, however, are useful. The challenge is to identify the (potential) interactions and to control them appropriately. Figure 4 schematically shows a typical situation in which resources are represented as boxes, and different workflows are sets of lines connecting them. Although the names of the systems have been removed to protect proprietary information, the picture represents the information system of a telecommunication company in the US.

The agent metaphor is useful when thinking about more than one work-flow. Agents can be identified not only with the passive resources on which the workflows execute, but also with the workflows themselves—the agents can then correspond either to the humans carrying out a given workflow, or the customer of a workflow. These agents must coordinate their efforts appropriately. For example, in a telecommunications setting, a channel assignment workflow must wait until enough channels have been created by another workflow.

The challenge is to learn the potential ordering constraints of the workflows. More generally, the challenge is to infer the activities or plans of other agents, and learn from repeated interactions with them. Related challenges arise when the information environment is truly open and new agents are added dynamically, or the agents involved do not repeat interactions. In such cases, an agent still needs to learn how to collaborate with classes of agents, and to classify them appropriately. For example, an agent may infer that agents who request a price quote for valves will often also want a price quote on matching hoses.

The foregoing challenges can be generalized still further to learn about the agents' dispositions to one another. For example, it is important to learn to what extent other agents will cooperate with the given agent. Indeed, if the agents form a team or coalition, they will be able to assist each other and prevent mishaps [Shehory & Kraus, 1996]. It is also useful to have models of the learning abilities of the other agents. A number of studies have shown that coalitions become more effective as the members of the coalition learn about each other. An implication of this is that the team members should act predictably and transparently (possibly by revealing their state) in order to abet the learning. Interestingly, this implication has not yet been researched or incorporated in any systems.

5 Abstractions and Structure

Learning about interactive components, such as roles and organizational structure, their dispositions, responsibilities, and commitments.

For CISs applied to enterprises or virtual enterprises, a variety of models are typically built. Figure 5 shows some of the common modeling approaches. Of the main ones, entity-relationship (E-R) diagrams describe a conceptual model of the information stored in (a subset of the databases in) the enterprise. Activity decomposition describes the relationship of inclusion among different activities, whereas the control, data, and materiel flows give additional information about it. E-R diagrams correspond to static information as in ontologies; the activity representations correspond to the workflows. It is important to relate the two categories of representations, because the actions in the workflows depend on the concepts they manipulate, and the concepts are defined based on their patterns of usage. A challenge is to classify the concepts and actions in this manner, so that they can be used for building ontologies and coordinating workflows.

In a number of settings, including enterprises, the organizational structure of a CIS is important. By the organizational structure, we mean the set of roles and responsibilities that make up a functioning system [Gasser, 1991; Papazoglou *et al.*, 1992]. There is an intimate relationship between the workflows executing in a CIS, and the organizational roles available in it. Figure 6 shows on the left a simple workflow corresponding to submitting a contract proposal from a company. The *write white paper* task itself may be decomposed into a subworkflow. The bottom left shows a possible subworkflow for travel. The tasks in the workflow impinge upon various databases, and other ongoing processes, such as *budget forecast*. They also relate to the organizational structure of the company, because key steps in the workflow must be performed by people with specific authorities.

Traditionally, the roles are mapped to tasks rigidly. However, in open and dynamic environments, more flexible role-bindings are needed. For example, if the *research director* is on leave, how may the workflow be rerouted? If one

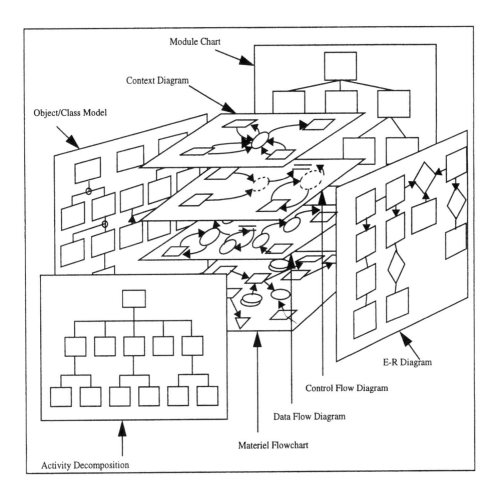

Fig. 5. Different Views of an Enterprise

person fills multiple roles, how may the workflow be scheduled to optimize their time? The challenge is to learn the capabilities and authorities necessary to execute different steps in a workflow, and to learn the interrelationships among the various roles.

6 Conclusions

Learning by agents can minimize or entirely replace communications, which is extremely important in large information environments where communication bandwidth is an expensive commodity.

In looking at the examples of CISs, we find that certain problems show up in different guises. The unifying themes of CIS are the following. One, we wish to obtain the effect of logical homogeneity and centralization despite physical distribution and heterogeneity. Two, we wish to support the logical openness of

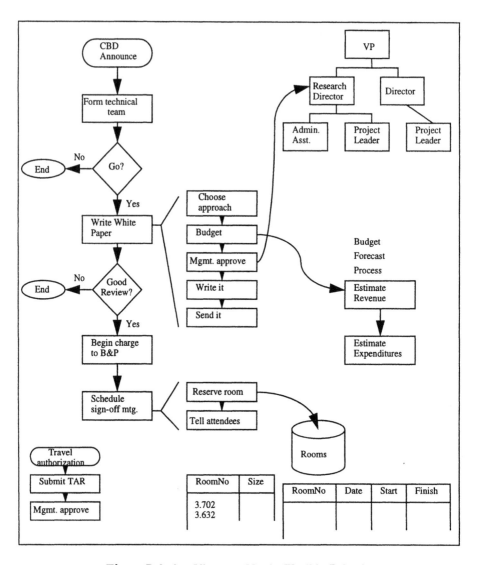

Fig. 6. Relating Views to Obtain Flexible Behavior

CISs. Openness translates into a number of interesting systemic challenges, relating to how a CIS may initialize and stabilize when some agents come together, are added, or leave. These lead to the following challenges for machine learning:

- learning about each other
- learning about society and the environment
- learning from repeat interactions with changing agent instances
- learning biased by social structure
- forgetting by a group about its former members.

A number of learning techniques exist [Russell & Norvig, 1995]. We give some suggestions about how different categories of learning might relate to problems in CIS. These categories are, of course, not mutually exclusive:

- Clustering techniques can help extract concepts from vast amounts of data (e.g., by classifying data that was carelessly mixed up)
- Passive learning appears appropriate for a new agent that joins a group (e.g., watching)
- Active learning can help a group learn about its new members, (e.g., interviewing them to evaluate their opinions)
- Unsupervised learning can be an unintrusive approach for acquiring workflows and learning the constraints on role-bindings (e.g., looking over the shoulder of staff members performing different tasks)
- Supervised learning applies for relating the more subtle interactions among workflows (e.g., being told business rules)
- Reinforcement learning applies in environments with autonomously built agents (e.g., adaptively acting and interacting).

This paper described some of the key ideas in CIS, and pointed out some of the places where machine learning could contribute. We believe the relationship between the two areas is synergistic. Cooperative information systems need machine learning to realize their promise of adaptivity and flexibility. Machine learning can benefit from CISs as a rich application area with open problems that are widely recognized as crucial, and promise to yield significant scientific advances in machine learning.

References

[Arens et al., 1993] Arens, Yigal; Chee, Chin Y.; Hsu, Chun-Nan; and Knoblock, Craig A.; 1993. Retrieving and integrating data from multiple information sources. *International Journal of Intelligent and Cooperative Information Systems* 2(3):127–158.

[Attie et al., 1993] Attie, Paul C.; Singh, Munindar P.; Sheth, Amit P.; and Rusinkiewicz, Marek; 1993. Specifying and enforcing intertask dependencies. In *Proceedings of the 19th VLDB Conference*. 134–145.

[Buchmann et al., 1992] Buchmann, Alejandro; Özsu, M. Tamer; Hornick, Mark; Georgakopoulos, Dimitrios; and Manola, Frank A.; 1992. A transaction model for active distributed object systems. In *[Elmagarmid, 1992]*. Chapter 5, 123–158.

[Bukhres & Elmagarmid, 1996] Bukhres, Omran A. and Elmagarmid, Ahmed K., editors. *Object-Oriented Multidatabase Systems: A Solution for Advanced Applications*. Prentice Hall.

[Chrysanthis & Ramamritham, 1994] Chrysanthis, Panos K. and Ramamritham, Krithi; 1994. Synthesis of extended transaction models using ACTA. *ACM Transactions on Database Systems* 19(3):450–491.

[Elmagarmid, 1992] Elmagarmid, Ahmed K., editor. *Database Transaction Models for Advanced Applications*. Morgan Kaufmann.

[Elmasri & Navathe, 1994] Elmasri, Ramez and Navathe, Shamkant; 1994. *Fundamental of Database Systems*. Benjamin Cummings, Redwood City, California, second edition.

[Fayyad, 1996] Fayyad, Usama, editor. *Special Issue on Data Mining*, volume 39(11) of *Communications of the ACM*.

[Gasser, 1991] Gasser, Les; 1991. Social conceptions of knowledge and action: DAI foundations and open systems semantics. *Artificial Intelligence* 47:107–138.

[Gray & Reuter, 1993] Gray, Jim and Reuter, Andreas; 1993. *Transaction Processing: Concepts and Techniques*. Morgan Kaufmann.

[Huhns & Stephens, 1989] Huhns, Michael N. and Stephens, Larry M.; 1989. Plausible inferencing using extended composition. In *Proceedings of the International Joint Conference on Artificial Intelligence*. 1420–1425.

[Huhns et al., 1994] Huhns, Michael N.; Singh, Munindar P.; Ksiezyk, Tomasz; and Jacobs, Nigel; 1994. Global information management via local autonomous agents. In *Proceedings of the 13th International Workshop on Distributed Artificial Intelligence*.

[Neches et al., 1991] Neches, Robert; Fikes, Richard; Finin, Tim; Gruber, Tom; Patil, Ramesh; Senator, Ted; and Swartout, William R.; 1991. Enabling technology for knowledge sharing. *AI Magazine* 12(3):36–56.

[Papazoglou et al., 1992] Papazoglou, Mike P.; Laufmann, Steven C.; and Sellis, Timothy K.; 1992. An organizational framework for cooperating intelligent information systems. *International Journal on Intelligent and Cooperative Information Systems* 1(1):169–202.

[Russell & Norvig, 1995] Russell, Stuart J. and Norvig, Peter; 1995. *Artificial Intelligence: A Modern Approach*. Prentice Hall, Upper Saddle River, NJ.

[Shehory & Kraus, 1996] Shehory, Onn and Kraus, Sarit; 1996. Formation of overlapping coalitions for precedence-ordered task execution among autonomous agents. In *Proceedings of the International Conference on Multiagent Systems*. 330–337.

[Shen & Leng, 1996] Shen, Wei-Min and Leng, Bing; 1996. A metapattern-based automated discovery loop for integrated data mining—unsupervised learning of relational patterns. *IEEE Transactions on Knowledge and Data Engineering* 8(6):898–910.

[Singh, 1996] Singh, Munindar P.; 1996. Synthesizing distributed constrained events from transactional workflow specifications. In *Proceedings of the 12th International Conference on Data Engineering (ICDE)*.

[Wiederhold, 1992] Wiederhold, Gio; 1992. Mediators in the architecture of future information systems. *IEEE Computer* 25(3):38–49.

[Zhang et al., 1996] Zhang, T.; Ramakrishnan, R.; and Livny, M.; 1996. BIRCH: An efficient data clustering mehtod for very large databases. In *Proceedings of the ACM SIGMOD Conference on Management of Data*.

A Modular Approach to Multi-agent Reinforcement Learning

Norihiko Ono and Kenji Fukumoto

Department of Information Science and Intelligent Systems
Faculty of Engineering, University of Tokushima
2-1 Minami-Josanjima, Tokushima 770, Japan

Abstract. Several attempts have been reported to let multiple mono-lithic reinforcement-learning agents synthesize coordinated decision poli-cies needed to accomplish their common goal effectively. Most of these straightforward reinforcement-learning approaches, however, scale poorly to more complex multi-agent learning problems, because the state space for each learning agent grows exponentially in the number of its part-ner agents engaged in the joint task. To remedy the exponentially large state space in multi-agent reinforcement learning, we previously proposed a modular approach and demonstrated its effectiveness through the ap-plication to a modified version of the pursuit problem. In this paper, the effectiveness of the proposed idea is further demonstrated using several variants of the pursuit problem. Just as in the previous case, our modu-lar Q-learning hunters can successfully capture a randomly-evading prey agent, by synthesizing and taking advantage of effective coordinated be-havior.

1 Introduction

In attempting to let simple reactive agents synthesize some coordinated behavior, several researchers in the fields of artificial life and machine learning have ap-plied monolithic reinforcement-learning algorithms to multi-agent learning prob-lems(e.g., [2, 5, 8, 9, 10]). In most of these applications, only a small number of learning agents are engaged in their joint tasks and accordingly the state space for each agent is relatively small. This is the reason why monolithic reinforcement-learning algorithms have been successfully applied to these multi-agent learning problems. However, these straightforward applications of reinforcement-learning algorithms do not successfully scale up to more complex multi-agent learning problems, where not a few learning agents are engaged in some coordinated tasks[9]. In such a multi-agent problem domain, agents should appropriately behave according to not only sensory information produced by the physical en-vironment itself but also that produced by other agents, and hence the state space for each reinforcement-learning agent grows exponentially in the number of agents operating in the same environment. Even simple multi-agent learn-ing problems are computationally intractable by the monolithic reinforcement-learning approaches.

Previously, to remedy the problem of combinatorial explosion in multi-agent reinforcement learning, we proposed a modular approach in [7], based on White-head's idea[12]. We considered a variant of the pursuit problem[1] as a multi-agent learning problem suffering from the combinatorial explosion, and showed how successfully modular Q-learning prey-pursuing agents synthesize coordinated decision policies needed to capture a randomly-moving prey agent. In this paper, we attempt to additionally validate the effectiveness of our modular approach through the application to other variants of the pursuit problem[1].

2 Problem Domain

In this paper, we consider a modified version of the pursuit problem[1] defined as follows:

> In an $n \times n$ toroidal grid world, a single *prey* and four *hunter* agents are placed at random positions in the grid, as shown in Fig. 1 (a). The hunters operate attempting to capture the randomly-fleeing prey. At every discrete time step, the prey and each hunter select their own actions independently without communicating with each other and accordingly perform them. Each agent has a repertoire of five actions. It can move in one of four principle directions (north, east, south, or west), or alternatively remain at the current position. The prey can not occupy the same position that a hunter does. However, more than one hunter can share the same position. The prey can not move to a position which has been occupied by a hunter, and vice versa. When the prey and a hunter attempt to move to the same empty position, both agents remain at their current positions. A hunter has a limited visual field of depth d $(2d + 1 < n)$. It accurately locates the relative position and recognizes the type(*hunter* or *prey*) of any other agent operating within its visual field, and it selects its next action according to the perceived information. The prey is captured, when all of its four neighbor positions are occupied by the hunters, as shown in Fig. 1 (b). Then all of the prey and hunter agents are relocated at new random positions in the grid world and the next trial starts. The ultimate goal for the hunters is to capture the prey as frequently as possible.

This is essentially the same problem as that considered in [7]. The only difference is that each hunter's identifier is *invisible* to any other hunters in this case, while it is *visible* in the previous one in [7].

3 A Modular Reinforcement Learning Approach

We attempt to let reinforcement-learning hunter agents spontaneously synthesize some collective behavior. As a basic learning algorithm, we have chosen to

---

[1] Preliminary results of this work have been reported in [6].

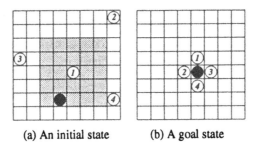

(a) An initial state   (b) A goal state

Fig. 1. The pursuit problem considered in this paper. Four hunters and a prey are indicated by white and black circles, respectively. Each hunter is assigned an identifier (1,2,3, or 4), but it is *invisible* to other hunters. A hunter is able to accurately recognize the relative position of any other hunter in its visual field of depth d. In the figure, the depth d equals 2, and the visual field of the hunter 1 is indicated by a shaded square in (a). The relative position of the prey with respect to the hunter 1 is $(-1, -2)$ in this initial configuration.

employ Q-learning. However, if we leave the agents' learning task to a standard monolithic Q-learning algorithm, we can have little hope to let them effectively learn to interact. Such a straightforward application of monolithic Q-learning to this kind of multi-agent learning problems brings about the problem of combinatorial explosion in their state spaces. Since multiple learning agents, each observing its partners' behavior, have to jointly perform a co-ordinated task, the state space for each learning agent grows exponentially in the number of its partners. For example, let us represent the state of a hunter agent by a combination of the relative positions of other agents, where the positions are represented by some unique symbols when they are not located within the hunter's visual field. Although there are only a small number of interacting agents involved, the size of state space for a hunter equals $4m^2 + 6(m^2 - 1)^2 + 2(m^2 - 1)(m^2 - 2)^2 + (m^2 - 1)(m^2 - 2)(m^2 - 3)^2/6$ where $m = 2d + 1$; the four terms in this expression stand for the cases when a total of 1, 2, 3, or 4 positions in the hunter's visual field are occupied by hunters[2]. The state space size is enormous even when the visual field depth d is small; e.g., at $d = 3$ and $d = 4$, it amounts to $1,021,700$ and $7,445,764$, respectively.

For this reason, we have chosen to implement each hunter agent by a variant of Whitehead's modular Q-learning architecture[12] as shown in Fig. 2. The architecture itself is identical to one we used to solve the pursuit problem in [7], but in this case we are forced to employ yet another modularization, because the previous one strongly relied on the assumption that each hunter's identifier is *visible* to any other hunters, but it does not hold here.

---

[2] Note that a hunter and the prey are not allowed to share the same position at a time and at least one position within a hunter's visual field is occupied by the hunter itself.

The architecture consists of three learning-modules and a single mediator-module. Each learning-module focuses on specific attributes of the current perceptual input and performs Q-learning. More specifically, at each time step, a hunter agent scans in the specific order, as shown in Fig. 3, individual positions within its visual field. Its i-th learning-module L_i receives only the relative position of the prey and that of the i-th partner that have been located during the scan[3], and ignores any information concerning the other partners[4].

Its state is represented by a combination of these two relative positions and the state space size equals $m^4 + 1$. This size of state space is computationally tractable by the monolithic Q-learning algorithm when d is relatively small. On the other hand, a mediator-module, without learning, combines learning-modules' decision policies using a simple heuristic decision procedure and selects a final decision by the corresponding agent. In our preliminary experiments, the mediator-module selects the following action:

$$arg \max_{a \in A} \sum_{i=1}^{3} Q_i(x_i, a)$$

where Q_i denotes the action-value function maintained by the learning-module L_i. This corresponds to what is called *greatest mass* merging strategy[12].

Our modular Q-learning architecture is based on the original one proposed in [12], but they are not identical. The architecture in [12] was particularly designed for constructing a *single* reactive planner pursuing a fixed set of dynamically-activated multiple goals. It was primarily employed for shrinking the planner's exponentially large state space caused by the existence of multiple goals. Here, our architecture is designed for shrinking each agent's exponentially large state space caused by the existence of its partners jointly working in the same environment. Note also that all the learning-modules, in our modular architecture, are always active and participate in the reinforcement-learning task unlike in the Whitehead's case. To use a set of dynamically activated learning-modules in this problem domain, we should modify the learning architecture so that a reward received by the learning agent propagates across the learning-modules[5].

4 Results

Our simulation run consists of a series of trials, each of which begins with a single prey and four hunter agents placed at random positions, and ends when the prey is captured. A trial is aborted when the prey is not captured within $1,500$ time

---

[3] When multiple partners occupy the same position, they are numbered randomly.

[4] Each learning-module in [7] only observes the relative position of the prey and that of a partner associated with some specific identifier. The previous modularization can not be employed here because each hunter is not able to recognize the located hunter's identifier.

[5] Note that in the Whitehead's modular architecture, any reward received by a learning-module does not propagate to the other learning-modules.

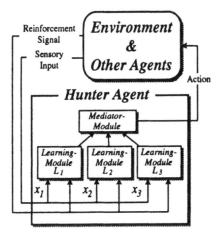

Fig. 2. Modular architecture for a hunter agent. At each time step, a hunter agent scans in a specific order individual positions within its visual field. Its component learning-module L_i only focuses on the prey and i-th partner that have been located during the scan, and does not concern any other partners' positions.

46	38	26	14	27	39	47
40	28	15	6	16	29	41
30	17	7	2	8	18	31
19	9	3	1	4	10	20
32	21	11	5	12	22	33
42	34	23	13	24	35	43
48	44	36	25	37	45	49

Fig. 3. Ordering of positions in each hunter's visual field. The visual field depth is supposed here to be 3. The 7×7 square indicates the hunter's visual field. The ordering reflects the Manhattan distance of each position to the hunter's position.

steps. Upon capturing the prey, individual hunters immediately receive a reward of 1.0, and accordingly all of its component learning-modules uniformly receive the same reward regardless of what decision policies they have. Our hunters receive a reward of -0.1 in any other case. The learning rate α and discounting factor γ are set to 0.1 and 0.9, respectively. Initial Q-values for individual state-action pairs are randomly selected from the interval $[0.01, 0.1]$.

We set each hunter's visual field depth to a small value of 3. By changing the dimension of the grid world, we observed how our learning hunter agents can reduce the average number of time steps to capture the prey. The results of our simulation runs are shown in Fig. 4. At initial trials, the hunters can not capture the prey effectively, but shortly they start improving their performance significantly. Eventually they come to steadily accomplish their goal at every trial in any runs of our simulation.

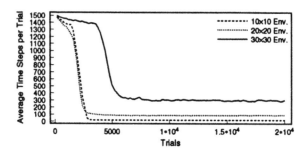

Fig. 4. Performance by modular Q-learning hunter agents. Four hunters, each with a fixed visual field depth of 3, are working together in environments with various dimensions. Each curve is an average over 25 simulation runs.

5   Synthesized Collective Behavior

Just as those modular Q-learning hunters in [7], our learning hunters exhibited two kinds of biologically-interesting collective behavior, namely, *functionality-specialization* and *herding*. Furthermore, some hunters exhibited *altruistic* behavior.

During the course of dramatically improving their performance, individual hunters are specializing their functionality. To capture the prey, individual hunters have to exclusively occupy one of four positions surrounding it. Let us call these positions *capture positions* of corresponding hunters. Eventually, our modular Q-learning hunters improve their pursuit performance by completely fixing their individual capture positions, i.e., by specializing their functionality. Fig. 5 illustrates how such a functionality-specialization is typically developed by hunters. Each of these four 3D line graphs shows a transition of the probability distribution over capture positions (indicated by relative directions to the prey) taken by the corresponding hunter. At initial 2,400 trials, hunters have not fixed their own capture positions, but after trial 2,500, they fixed their own capture positions and during this change in their behavioral patterns, their overall performance was dramatically improved.

Besides specializing their functionality, our hunters obtain a kind of herding behavior. Even if having established the above-mentioned static role-assignment, hunters can not always accomplish the capturing task effectively until they have obtained an effective way of collectively locating the prey, since each of them can only observe a limited portion of the grid world.

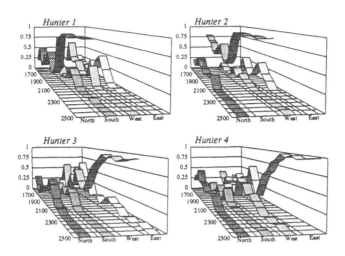

Fig. 5. Typical process of specialization.

Our learning hunters do not attempt to locate the prey independently, because it is not an effective way even when the prey behaves randomly. Instead they attempt to locate it in a coordinated manner. Once having located a partner, a hunter attempts to keep the partner in sight so long as the prey has not been located and thereby it makes a herd together with the partner. By herding in this way, hunters can compensate their limited perceptual capabilities and operate as if in larger visual fields, as shown in Fig. 6 (a). Their behavior suddenly changes once one of the herding members has happened to locate the prey. The member quickly attempts to reach its capture position and the other ones follow it and accordingly are likely to locate the prey. When all of hunters have located the prey, they can easily capture it.

In his simplified version of the pursuit problem, Tan[9] allowed his Q-learning hunters to share their perceptual inputs and attempted to let them acquire *mutually-scouting* behavior. Although the herding behavior automatically synthesized by our hunters is not identical to the mutually-scouting behavior, both behavior has similar effects. Note that we did not explicitly intend to let our hunters synthesize herding behavior, while Tan intended to explicitly let his hunters share information and acquire scouting behavior.

In addition to the above-mentioned collective behavior, some of our learning hunter often exhibits a kind of *altruistic* behavior. When the hunter locates the prey but not all of the other hunters are visible within its visual field, it does not attempt to reach its responsible capture position immediately. Instead, as illustrated in Fig. 6 (b), the hunter keeps a certain distance from the prey so that it becomes easier for the other hunters to locate the hunter itself or the prey. The same behavior was also exhibited by the learning hunters in [7], which was not mentioned in [7].

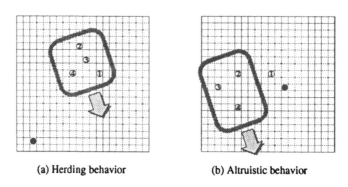

(a) Herding behavior   (b) Altruistic behavior

Fig. 6. Herding and altruistic behavior by modular Q-learning hunters. Each hunter's visual field depth is supposed to be 3 here. By making a herd and moving in a consistent direction(indicated by an arrow), the hunters can operate as if in larger visual fields and effectively search the prey, as suggested in (a). The hunter 1 behaves altruistically in the situation (b) where the hunter has located the prey but no other hunters are visible. By keeping a certain distance from the prey, it attempts to assist the collective search by the other hunters.

Typically, the prey capturing process by our modular Q-learning hunters consists of two phases: (i) a *collective search phase* where all of hunters attempt to identify the prey's position by configuring a herd, and (ii) a *surrounding phase* where each hunter having identified the prey's position attempts to reach its fixed capture position in an effective way.

To illustrate this coordinated behavior, a typical prey pursuit process by our learning hunters is shown in Fig. 7. The hunters have already experienced some 10,000 trials in a 20 × 20 toroidal environment, with a visual field depth of 3 and a randomly evading prey.

So far we have supposed that every hunter has the same visual field uniformly. There is no especially effective specialization among the 24 possibilities, and any of them may be developed eventually by the learning hunters. Thus, we can not foresee which specialization will be eventually developed prior to a simulation run.

What will happen if the hunters have nonuniform visual fields? Are our learning hunters able to steadily identify an appropriate specialization? We consider one of the most interesting cases where the learning hunters have different visual fields as shown in Fig. 8. In this case, our modular Q-learning hunters have developed the same specialization very quickly at every simulation run we performed. A typical specialization process by the hunters is depicted in Fig. 9. Note that they agreed on the eventual specialization much more quickly than the hunters with uniform visual fields.

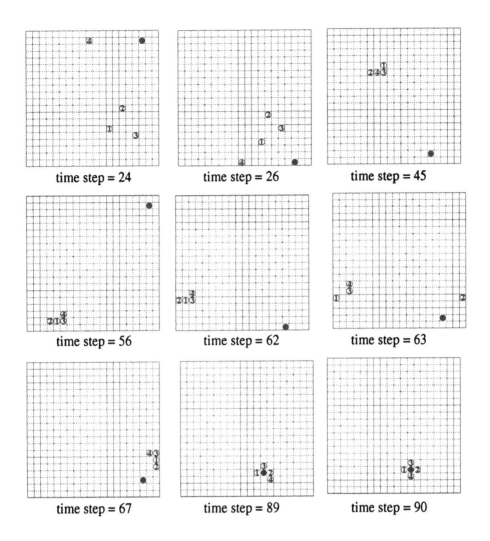

Fig. 7. Typical pursuit process by hunters.

6   Other Cases

In any variants of the pursuit problem we considered above, the prey agent is supposed to evade randomly. This assumption might make the problem too easy to solve. It has not been clear how effectively our modular Q-learning hunter agents can capture the prey employing other evading strategies.

We investigated the performance by our learning hunters especially when the prey moves linearly or escapes from the located hunters. More specifically, we

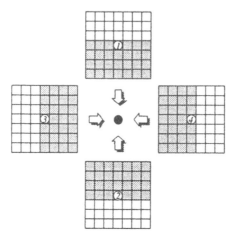

Fig. 8. Four hunter agents with nonuniform visual fields. A hunter is indicated by a white circle, whose visual field is depicted by a shaded rectangle surrounding it. At each simulation run, these hunters quickly specialize their functionality and eventually they come to capture the prey, indicated by a black circle, from the direction indicated by their corresponding arrows.

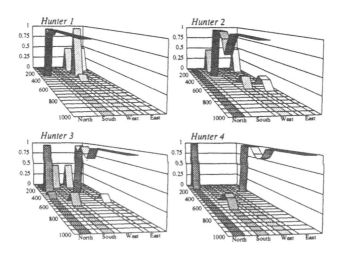

Fig. 9. Typical specialization process by hunters with nonuniform visual fields.

simulated the behavior of modular Q-learning hunters in the following variants of the pursuit problem:

1. The prey has two possible actions only; it can move east or remain at the current position.
2. The prey has a repertoire of three actions only; it can move north and east, or alternatively stand still.
3. The prey evades randomly when no hunter is located, and otherwise it selects its actions so as to maximize the Manhattan-distance to the nearest hunter within its visual field. The prey is supposed to have the same visual field depth as that of a hunter.
4. The prey is supposed to have a visual field of limited depth. It evades randomly when no hunter is located, and otherwise it behaves so as to maximize the total sum of Manhattan-distances to the located hunters.

A typical performance by our learning hunters in cases 1–4 are shown in Fig. 10 where all the hunters have the same visual field depth of 3. The last two cases are difficult to solve using our modular approach. Our hunters take considerable time to synthesize effective behavior in case 3. In case 4, it is almost impossible for our learning hunters to get the first positive reward, if the visual field depth of the prey equals that of a hunter. However, if the visual field depth of the prey is shallower than that of the hunters, it is possible for our hunters to synthesize effective coordinated behavior to capture the prey, as suggested in Fig. 10 (d) where the prey has a visual field depth of 2.

So far it has been supposed that the hunters can share their positions. This assumption also may make the problem too easy to solve. We investigated the performance by our modular Q-learning hunters, provided that multiple hunters are not allowed to occupy the same position. Typical simulation results are shown in Fig. 11. Each curve in the figure is an average over 10 simulation runs. The hunters take substantial time to acquire appropriate coordinated behavior, but eventually they come to effectively capture the escaping prey. A typical pursuit process by the hunters is also shown in Fig. 12.

7 Concluding Remarks

Recent attempts to let monolithic reinforcement-learning agents synthesize some of coordinated relationships among them scale poorly to more complicated multi-agent learning problems where multiple learning agents play different roles and work together for the accomplishment of their common goals. These learning agents have to receive and respond to various sensory information from their partners as well as that from the physical environment itself, and hence their state spaces are subject to grow exponentially in the number of the partners.

To remedy the problem of combinatorial explosion in multi-agent reinforcement learning, we have proposed a modular approach in [7]. In this paper, we additionally investigated the effectiveness of the idea. As illustrative problems

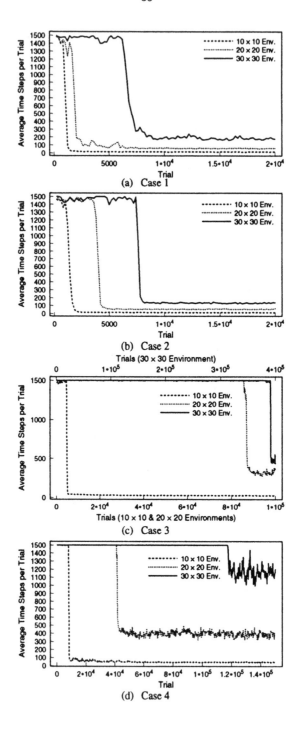

Fig. 10. Performance by learning hunters pursuing the prey employing various evading strategies.

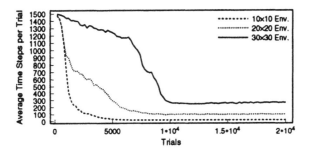

Fig. 11. Performance by learning hunters pursuing an randomly evading prey. More than one agent can not occupy the same position. The hunters have the same visual field depth of 3.

suffering from the combinatorial explosion, we considered several variants of the pursuit problem. We showed how successfully a collection of newly designed modular Q-learning hunter agents can capture a randomly-evading prey agent, by specializing their functionality and synthesizing herding and altruistic behavior.

Multi-agent learning is a difficult problem in general, and the results we obtained strongly rely on specific attributes of the problems just as in our previous case[7]. But the results are quite encouraging and suggest that our modular reinforcement-learning approach is promising in studying adaptive behavior of multiple autonomous agents. In the next phase of this work we intend to investigate the effects of the modular approach on other types of multi-agent learning problems.

References

1. Benda, M., V.Jagannathan, and R.Dodhiawalla: On Optimal Cooperation of Knowledge Sources, Technical Report BCS-G2010-28, Boeing AI Center, 1985.
2. Drogoul, A., J.Ferber, B.Corbara, and D.Fresneau: A Behavioral Simulation Model for the Study of Emergent Social Structures, F.J.Varela, *et al.* (Eds.): Toward a Practice of Autonomous Systems: Proc. of the First European Conference on Artificial Life, The MIT Press, 1991.
3. Gasser, L. *et al.*: Representing and Using Organizational Knowledge in Distributed AI Systems, L.Gasser, and M.N.Huhns (Eds.): Distributed Artificial Intelligence, Vol.II, Morgan Kaufmann Publishers, Inc., 1989.
4. Levy, R., and J.S.Rosenschein: A Game Theoretic Approach to Distributed Artificial Intelligence, MAAMAW'94 Pre-Proc. of the 3rd European Workshop on Modeling Autonomous Agents in a Multi-Agent World (available as technical document D-91-10 of German Research Center on AI), 1991.
5. Ono, N., T.Ohira, and A.T.Rahmani: Emergent Organization of Interspecies Communication in Q-learning Artificial Organisms, in F.Móran *et al.*: (Eds.) Advances in Artificial Life: Proc. of the 3rd European Conference on Artificial Life, Springer, 1995.

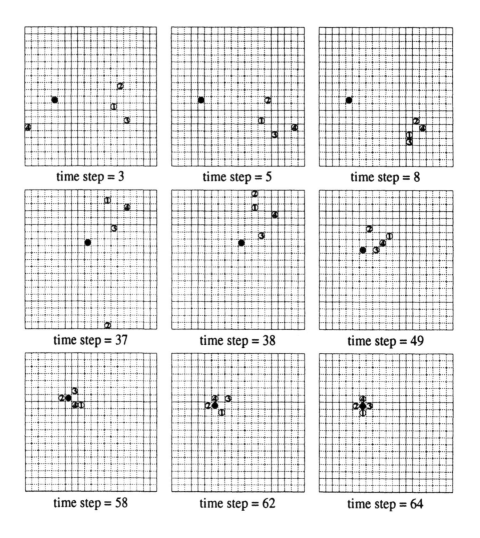

Fig. 12. Typical pursuit process by hunters. The agents can not share their positions.

6. Ono, N., and K.Fukumoto: Collective Behavior by Modular Reinforcement-Learning Animats, P.Maes *et al.*(Eds.): From Animals to Animats *4*: Proc. of the 4th International Conference on Simulation of Adaptive Behavior, The MIT Press, 1996.

7. Ono, N., and K.Fukumoto: Multi-agent Reinforcement Learning: A Modular Approach, Proc. of the 2nd International Conference on Multi-agent Systems, AAAI Press, 1996.

8. Rahmani, A.T., and N.Ono: Co-Evolution of Communication in Artificial Organisms, Proc. of the 12th International Workshop on Distributed Artificial Intelligence, 1993.

9. Tan, M.: Multi-agent Reinforcement Learning: Independent vs. Cooperative Agents, Proc. of the 10th International Conference on Machine Learning, 1993.

10. Yanco, H., and L.A.Stein: An Adaptive Communication Protocol for Cooperating Mobile Robots, From Animals to Animats 2, The MIT Press, 1992.

11. Watkins, C.J.C.H.: Learning With Delayed Rewards, Ph.D.thesis, Cambridge University, 1989.

12. Whitehead, S. et al.: Learning Multiple Goal Behavior via Task Decomposition and Dynamic Policy Merging, in J.H.Connell et al. (Eds.): Robot Learning, Kluwer Academic Press, 1993.

Learning Real Team Solutions

Cristina Versino and Luca Maria Gambardella

IDSIA, Corso Elvezia 36, CH-6900 Lugano, Switzerland
{cristina,luca}@idsia.ch, http://www.idsia.ch

Abstract. This paper presents "Ibots" (Integrating roBOTS), a computer experiment in group learning designed on an artificial mission. By this experiment, our aim is to understand how to use reinforcement learning to program automatically a team of robots with a shared mission. Moreover, we are interested in *learning real team solutions*. These are programs whose form strongly depends on the number of robots composing the team, on their individual skills and limitations, and on any other mission boundary condition which makes it worth to prefer "at a team level" certain solutions to others. The Ibots mission is specified implicitly by means of a single reinforcement signal which measures the team performance as a whole. This form of payoff leads to real team solutions. Benefits and drawbacks of using team reinforcement as opposed to individual robot reinforcement are discussed.

1 Introduction

The use of reinforcement learning [1] to produce self-programming robots is not new in the context of *single robot* missions, where a reinforcement signal directly evaluates the behavior of the only robot in charge of carrying out the task.

A Robot Credit Assignment Problem. The picture changes as *many robots* are acting at the same time, with little or perhaps no knowledge at all about teammate activities. In this scenario, if the reinforcement signal reflects the whole team performance, each single robot is faced with the problem of deciding to what extent its own behavior has contributed to the overall team's good or bad score: this is the *robot credit assignment problem*[1]. Because of the robot credit assignment problem, each robot has a *noisy* perception of the mission it is asked to accomplish. A single robot can behave identically many times (during different trials of the mission), and nevertheless, it may receive completely different payoffs. This occurs because it is not the only actor, and the reinforcement signal just *partially* depends on its actions.

Bypassing the Robot Credit Assignment Problem. Instead of addressing the robot credit assignment problem directly, one can bypass it by reformulating the team learning problem.

A *first* way is to enable *broadcast communication* between teammates. If a robot is *aware* of other robots' perceptions and actions, then it is in a position to

---

[1] In [2], this problem is called *inter-agent credit-assignment problem*.

make sense out of a global team payoff. Explicit communication makes the team equivalent to a "big robot" [3], whose perception and action are the union of perceptions and actions of all team members. Seen in this way, the team payoff is the measure of the big robot performance. The utility of communication has been proved experimentally for small teams of real robots both in a simplified hazardous waste cleanup mission [4], and in a box-pushing task [5, 6]. However, communication is not always possible technically, and it tends to become a bottleneck as the team size increases [7].

A *second* way of avoiding the robot credit assignment problem is to measure *each robot individual performance* instead of team performance. In [8, 9] this idea is applied to training a group of real robots in a foraging task. Pucks disseminated in the workspace have to be collected and delivered to a home area. Each robot in the team learns a personal policy through individual payoff. For example, a robot is rewarded whenever it grasps a puck or if it drops a puck at home. In this framework, a single robot is not interested in the performance of its teammates, because it addresses the mission in an individualistic sense. We see *two drawbacks* in this approach. *First:* its underlying assumption is that team performance indirectly increases because individual performance increases. However, if the robots do not learn the task at a *similar pace*, it cannot be guaranteed that each robot will learn and participate to the mission. If not all robots learn how to contribute to the mission, the team performance will be suboptimal. As an example, suppose that in the foraging task one robot in the team manages to learn (maybe just by chance) the individually optimal policy after a few trials. This "superrobot" will collect most of the pucks by itself, diminishing the learning opportunities of its teammates because pucks are a *limited, shared resource*. To improve this state of affairs, the superrobot should behave in a suboptimal way (by forgetting its optimal policy) so as to let the other robots take part in the task and learn. However, there is no reason for the superrobot to recede from its optimal policy, because it is designed to be the best possible individual. The superrobot phenomenon is not as unrealistic as it appears at first glance. It may arise, for instance, in *incremental learning* experiments where the team size is progressively increased: elder, experienced robots would tend to carry out the mission by themselves, limiting the possibility of novice robots to participate and learn. In [10], it is argued that *social rules* can be learnt by the robots to minimize resource competition and to direct their behavior away from individual greediness and towards global efficiency. The idea is interesting, but the experimental results reported in [10] are preliminary. The *second* drawback we find in the individualistic approach is summarized in the answer to the following question. Suppose that, in a team of homogeneous robots, all robots receive personal reinforcement signals generated by the same payoff function: where are the policies learnt by the robots expected to converge? In general, to the optimal policy for a robot carrying out the mission by itself. The robots will behave as "clones" of *a robot designed to work alone*.

Real Team Solutions. We feel this is not the spirit of group learning, which should be aimed at producing *efficient real team solutions. These are policies whose form is strongly influenced by the number of robots in the team, by each robot's skill and weakness, and by any other mission boundary condition which is relevant for preferring "at a team level" some solutions to others.* Real team solutions encourage the participation of all robots which are in a position to positively contribute to the mission.

Facing the Robot Credit Assignment Problem. *To obtain truly team solutions, one should use team payoffs at the price of dealing with the ambiguity posed by the robot credit assignment problem.* Multi-agent learning experiments based on team payoffs are illustrated in recent works [11, 12, 13, 14].

In [11] a team of simulated agents learns signaling behaviors to efficiently solve an object-gathering task in an unknown and changing environment. The reinforcement signal is based on the total time needed by the team to gather all the objects in the workspace. Experiments are carried out under several conditions: with teams of different size, with a variable number of objects, and with different object distributions; moreover, during learning the object distribution is occasionally modified to test the agents ability of traking environmental changes. Under these heterogeneous experimental conditions, each agent learns the appropriateness of exhibiting a given signaling behavior. For example: an agent perceiving an object is faced with the problem of deciding whether to activate its "object-signaling" behavior to attract other agents towards its position. The agent learns that exhibiting this signaling behavior is, in general, rewarding for the team when the objects are distributed in clusters: in this case, the detection of one object gives a high chance of finding other objects in the neighborhood, and these could be collected by the called agents. On the contrary, the same signaling behavior is inappropriate if the objects are uniformly distributed in the environment: the agent should refrain from calling other teammates whenever it finds an object. Finally, certain mission boundary conditions require the agents to acquire different signaling policies: they specialize into *signallers* and *harvesters*. The authors show by statistical analysis of the results that the team discovers by trial-and-error a near-to-optimal signaling policy given the specific mission conditions.

In [12] a team of Q-learning agents is engaged in the challenging real-world problem of elevator dispatching. Each agent is responsible for controlling one elevator car. Two different control architectures are tested. In the *parallel* architecture, the agents share a single neural network which models a common policy: this allows the agents to learn from each others experiences but forces them to use identical policies. In the *decentralized* architecture, the agents learn personal networks, which allow them to specialize their control policies. The team receives as global payoff the sum of squared wait times of passengers. Despite this noisy reinforcement signal and the inherent stochastic nature of the task, results obtained in simulation on both architectures surpass the best known heuristic elevator control algorithm. The authors expect an additional advantage of reinforcement learning over heuristic controllers in buildings with heterogeneous arrival rates at each floor, because Q-learning agents may adapt to each floor traffic profile.

In [13] a general method for incremental self-improvement and multi-agent learning in unrestricted environments is presented. In one of the implementations, a recurrent neural net is applied to a non-Markovian maze task. Each connection in the net is viewed as an agent: the connection's weight represents the agent's policy. The net learns to guide an animat to a goal by using a team payoff whose value turns from 0 to 1 only once the animat hits the goal. Looking at the net's connections as if they were a team of agents is an unconventional point of view. But why not doing so? After all, a neural net is a good example of a set of partially independent agents (the net's weights, or the neurons themselves) which learn to act well "as a team". Following this view, the structural credit assignment problem in connectionist reinforcement learning can be regarded as being equivalent to the robot credit assignment problem described above.

Finally, this paper presents "Ibots" (Integrating roBOTS) [14], an experiment in group learning designed to understand how to use reinforcement learning to program automatically a team of robots with a shared mission. As in [11, 12, 13], Ibots learn through a reinforcement signal which measures the team performance as a whole. In this way, Ibots manage to learn real team solutions.

2 General Issues

Before describing the Ibots mission, we summarize the general issues addressed by the experiment, as well as the underlying assumptions and design choices.

Team Size. In the Ibots experiment, the same mission is handled with teams of *different size*. By this, our aim is to assess whether learnt policies change as the team size changes.

Team Composition. Ibots can be *homogeneous* or *heterogeneous*. When they are homogeneous, they all have same sensors and same actuators. When they are heterogeneous, they have different sensors and/or different actuators. When Ibots are heterogeneous, they have potentially different skills and weaknesses due to physical characteristics. The question is whether they can learn to specialize their behavior so as to emphasize skills and minimize the impact of weaknesses. The same question is addressed in [15].

Mission's Boundary Conditions. Ibots are confronted with *different boundary conditions* to the mission. We are interested in checking whether the strategy learnt by the team to accomplish the mission changes as the boundary conditions change. [11] also address this question by dealing with different object distributions in their object-gathering task.

Control Programs. Ibots run and learn control programs defining their behavior in the mission. The Ibots' control programs may be *public* or *private*. When they are public, all Ibots share the same program; when they are private, each Ibot works with a different, personal program.

Robots working with private programs instead of public programs is more general. If we choose the public program option, we assume that there exists a shared solution which works for every robot in the team. This is not the case when the robots are heterogeneous or when the mission requires the robots to specialize their behavior. If the robots are homogeneous and the mission does not require specialization, it makes sense to consider the public option, because it is faster to learn 1 shared program than n distinct programs, n being the number of robots in the team. Learning a shared program is faster for two reasons: first, the search space for the learning algorithm gets reduced, and, second, the robot credit assignment problem completely vanishes. Sometimes, even when the mission requires the robots to specialize their behaviors, it is possible to define the team learning problem in such a way that the public program option is still valid. As an example, we guess that if the agents in [11] were designed to learn directly the distribution of object-signaling agents over the whole team (instead of the individual agent's tendency to exhibit the object-signaling behavior), it would be easier for the team to learn the correct proportion between signaling and non-signaling agents. Finally, we stress that learning 1 shared program in a team of n robots is not the same as learning 1 program with a single robot and then cloning it n times. *A shared program is a real team solution, while cloned programs are not.* The Ibots experiment illustrates this point clearly (see "Public Control Programs", in Section 6.2).

In our opinion, the possibility of learning a single shared program instead of several private programs has been overlooked by most collective robotics works. This is curious if one considers the recent trend of designing robot teams inspired to Swarms [16]. Robot Swarms have shown that an interesting group behavior may emerge from the interaction between robots running a public control program. Odd enough, when learning techniques are used to derive automatically a team behavior, people usually neglect the possibility of learning a public program.

The only work we are aware of where both public and private policies have been considered, is the recent paper on elevator dispatching by [12]. Their "parallel" and "decentralized" architectures are equivalent to our "public" and "private" policies, respectively. In this application, however, it remains unclear which are the benefits of using the decentralized architecture in a problem where the elevator cars to be controlled are homogeneous.

Communication. Ibots do *not communicate*. By this we mean that, when Ibots work with private control programs, they do not tell each other which these programs are. In this restricted sense, there is no communication. By this choice we depart from works in collective robotics where robots are aware of teammates activities through explicit communication [5, 4, 15, 6].

Sharing Limited Resources. In collective robotics, robots usually share physical resources (the workspace, objects to gather, etc).

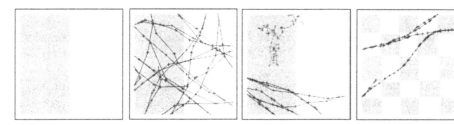

Fig. 1. (From left to right) (a) The squared arena and the "half-full" *Region*: $I = 0.49$. The arena side is 500 unit long. (b) Trajectory of one Ibot running a solitary trial with $N_{max} = 100$ and $Prog = (30°, 200)$. Crosses indicate sampled points. The result is $N_{in} = 52$, $\hat{I} = 0.52$. (c) Trajectories of two Ibots running a shared trial with $N_{max} = 100$, starting from a scattered configuration and running private control programs: $Prog^1 = (180°, 50)$ (gray trace), $Prog^2 = (20°, 150)$ (black trace). The result is $N_{in}^1 = 55$, $N_{in}^2 = 30$, $\hat{I} = 0.85$. Ibot 1 took $N_{sam}^1 = 61$ samples, Ibot 2 took $N_{sam}^2 = 39$. (d) Trajectory of one Ibot running a trial on the "chessboard" *Region*: $I = 0.49$. $N_{max} = 100$ and $Prog = (10°, 40)$. The result is $N_{in} = 45$, $\hat{I} = 0.45$.

In our experiment, Ibots share a limited, non-physical resource. However, there is no a priori rule which decides how this resource should be shared. We thought the Ibots should be more or less resource-*greedy* depending on their control programs. Thus, while the Ibots learn their control programs, they implicitly learn to share the resource in a way which is convenient for the team. In particular, when they run *public programs*, they exhibit the same greediness, and, on the average, each Ibot takes the same amount of the total resource. This is not the case when the Ibots run *private programs*, where each Ibot has a different propensity in consuming the resource. This fact has a great impact on the team performance in the mission.

Learning through Reinforcement. Ibots learn to accomplish the mission through *trial-and-error*. At each trial, each Ibot receives a *reinforcement signal* which measures the *team performance*. Ibots are confronted with their performance as a group because we want them to learn a team strategy. This view is shared with [11, 12, 13], but it is different from that pursuited in [8, 9], where learning is driven by individual performance.

3 Ibots

The *mission* for the Ibots is to guess the integral I $(0 \leq I \leq 1)$ of an arbitrary gray *Region* drawn on the white ground of a squared arena (Fig. 1(a),(d)).

How are robots turned into Ibots? For the sake of clarity, let us first consider the case of a team composed of one Ibot.

3.1   One Ibot

The Ibot's dowry is a *control program Prog* which lets it explore the arena while sampling the ground color. By activating *Prog*, the Ibot performs a *trial* run (Fig. 1 (b),(d)).

A trial starts from a random location in the arena. It is a sequence of N_{max} *elementary movements* separated by stops. Whenever the Ibot stops, it samples the ground color. A color reading equal to gray gives evidence for the sample to be "inside *Region*", while a white reading is interpreted as "outside *Region*". At the end of the trial, the Ibot returns the number N_{in} of samples it counted inside *Region*. $\hat{I} = N_{in}/N_{max}$ is its estimate of I at this trial. $E = |I - \hat{I}|$ is the error in the estimate.

How are elementary movements generated? The Ibot control program *Prog* depends stochastically on two parameters $(\alpha_{prog}, \delta_{prog})$ which remain fixed during a trial. α_{prog} is used to generate a rotation instruction for the Ibot, while δ_{prog} induces a translation. The semantics of α_{prog} and δ_{prog} is as follows. First, a number α is drawn from a uniform distribution in $[-\alpha_{prog}, +\alpha_{prog}]$. This is interpreted by the Ibot as: "Rotate α degrees.". Second, a number δ is drawn from a uniform distribution in $[0, \delta_{prog}]$. The interpretation for δ is: "Translate δ units in your current heading direction, calling the *bumping rule* if necessary.". The bumping rule is called when the Ibot meets the arena border before having covered the whole distance δ. In this case, the Ibot rotates $180°$ and covers the remaining distance *minus* 1: bumping against the arena border consumes 1 unit of translation to prevent the Ibot from bumping forever without consuming δ, a rare case which may occur when the Ibot lies in a corner. The bumping rule can be called recursively. We decided the bumping rule should just reverse the Ibot's travelling direction to affect minimally its natural motion angle which already depends on the parameter α_{prog}.

In our computer experiment, a rotation instruction is executed by the Ibot in one time unit whatever the rotation angle α; while the execution time of a translation instruction is directly proportional to the distance δ.

Finally, both program parameters α_{prog} and δ_{prog} take values in a finite range: $0 \leq \alpha_{prog} \leq \alpha_{max}$ and $0 \leq \delta_{prog} \leq \delta_{max}$.

3.2   A Team of Ibots

Each Ibot i $(i = 1, \ldots, N_{ibots})$ being equipped with a control program $Prog^i = (\alpha^i_{prog}, \delta^i_{prog})$, there are several ways of generalizing from the single Ibot case to the team case (Fig. 1 (c)). We have considered both the cases where the $Prog^i$s may be *public* or *private*. Public means that all instances of $Prog^i$s are constrained to be the same $Prog$, while private means that each $Prog^i$ may be different.

The Ibots activate the $Prog^i$s in parallel to run a team trial. At the beginning of the trial, the Ibot locations are chosen at random in the arena, and these are either *clustered* or *scattered*. In the clustered configuration, all Ibots have the same initial position and orientation; in the scattered configuration, they have different positions and orientations. During a trial, the Ibots are granted a total of N_{max} elementary movements to collect N_{max} samples of the ground color *overall:* *samples are the team limited and shared resource.* Whenever an Ibot stops to take a sample, it is allowed to do so only if less than N_{max} samples have been taken by the team so far. Otherwise, the Ibot gives up the sampling, and the team trial is terminated. Notice that when the Ibots work with private *Prog*^{i}s, each Ibot i will collect a different number of samples N_{sam}^i: Ibots with small δ_{prog}^is (travelling for shorter distances) will take on the average more samples than Ibots with larger δ_{prog}^is. At the end of the trial, each Ibot i returns the number N_{in}^i of samples it counted inside *Region*. These contributions are summed in $N_{in} = \sum_i N_{in}^i$, leading to the team integral estimate $\hat{I} = N_{in}/N_{max}$, the error being $E = |I - \hat{I}|$.

Finally, Ibots are *immaterial,* they do not collide when their trajectories intersect.

4 Programmed Ibots vs. Learning Ibots

How can a team of Ibots learn to provide good estimates $\hat{I}$ of the integral I? *The goal of group learning is to find control programs Prog^{i}s leading to estimates $\hat{I}$ close to I.* As the behavior of each program depends on its parameters α_{prog}^i and δ_{prog}^i, *the target of learning is to discover "good" pairs $(\alpha_{prog}^i, \delta_{prog}^i)$.* Notice that we know a *general* solution, namely:

$$\forall i : \begin{cases} \alpha_{prog}^i = 180° \\ \delta_{prog}^i = \text{"length of arena diagonal"} \end{cases} \tag{1}$$

With this choice of parameters, an Ibot may reach *any* position in the arena starting from *any* other position and orientation in just *one* elementary movement. Knowing the Ibot's current position and orientation, its next position remains highly unpredictable. An observer would describe the points sampled by this Ibot as being *uniformly distributed* in the arena. This brings us to the hypothesis of the Monte Carlo method [17] for integration. This states that, by drawing N_{max} points from a uniform distribution in the arena, the error E in the estimate of the integral is probabilistically bounded by N_{max}:

$$P\left\{E \leq \frac{1}{\sqrt{N_{max}}}\right\} \geq 0.9999 \tag{2}$$

For example, by drawing 100 points, we are almost guaranteed that the error in the estimate will not exceed 0.1, whatever the *Region*'s integral and shape (remember that $0 \leq I \leq 1$). Given N_{max}, this result quantifies the *admissible error* for the Ibots mission. We call the control programs defined by Eq. 1 the "programmer solution", because it reflects, in our opinion, the way a programmer would address this robot programming task: by looking for a *general* solution, which will work for any *Region*, whatever the number of Ibots in the team, independently of their starting configuration in the arena. Though appealing, we are not interested in this *a priori* solution. Rather, we are looking for *real team solutions* established through experience. These should depend on the specific *Region*, on the number of Ibots, on their initial configuration, and on their specific skills when these latter are no longer homogeneous.

To see why this "adaptation to circumstances" makes sense, consider a team of 100 Ibots started in a scattered configuration, i.e. each Ibot's initial position is drawn at random in the arena. In this situation, it would be *unnatural* to see the Ibots running the programmer's solution, as the integral is perfectly guessable by the more economic "staying in place" program:

$$\forall i : \begin{cases} \alpha^i_{prog} = 0° \\ \delta^i_{prog} = 0 \end{cases}$$

How can real team solutions be derived? When dealing with private control programs, exaustive search in the multi-dimensional program space of the Ibots is unfeasible. Suppose the Ibot's *continuous* program space is discretized so that each Ibot selects its control program only among a set of N_{progs} different control programs. In a team of size N_{ibots}, the number of possible distinct program selections for the team is [18]:

$$\binom{N_{ibots} + N_{progs} - 1}{N_{ibots}} = \frac{(N_{ibots} + N_{progs} - 1)!}{N_{ibots}!(N_{progs} - 1)!}$$

Moreover, each program selection should be tested several times to be sure about its quality, because "bad" control programs have a stochastic performance. As an alternative to exaustive search, true team solutions can be derived through learning: by measuring the Ibots team performance as a whole.

5   Learning to Be Good Ibots

"Good" *Prog^i*s are programs which lead *repeatedly to admissible* estimates of I. Admissible estimates are defined with respect to N_{max} by Eq. 2. Moreover, estimates must be repeatable because we are interested in programs with a *stable* performance.

The Ibots learn good control programs *Prog^i*s by reinforcement through a sequence of trials. Each time instant t corresponds to a team trial. Let $Prog^i(t) = (\alpha^i_{prog}(t), \delta^i_{prog}(t))$ be Ibot i control program at time t. The following 4 steps are repeated forever.

1. *Each Ibot "i" independently generates a new, tentative control program* $New^i(t) = (\alpha^i_{new}(t), \delta^i_{new}(t))$ *by slightly modifying* $Prog^i(t)$.

Given:

$$\alpha^i_{temp} = \alpha^i_{prog}(t) + \alpha^i_{rand}(t) \cdot \alpha_{step}$$
$$\delta^i_{temp} = \delta^i_{prog}(t) + \delta^i_{rand}(t) \cdot \delta_{step}$$

it is:

$$\alpha^i_{new}(t) = \begin{cases} 0° & \text{if } \alpha^i_{temp} < 0° \\ \alpha_{max} & \text{if } \alpha^i_{temp} > \alpha_{max} \\ \alpha^i_{temp} & \text{otherwise} \end{cases}$$

$$\delta^i_{new}(t) = \begin{cases} 0 & \text{if } \delta^i_{temp} < 0 \\ \delta_{max} & \text{if } \delta^i_{temp} > \delta_{max} \\ \delta^i_{temp} & \text{otherwise} \end{cases}$$

where:

- $\alpha^i_{rand}(t)$ and $\delta^i_{rand}(t)$ are uniform random numbers in $[-\rho(t), +\rho(t)]$ (see below for the definition of $\rho(t)$);
- α_{step} and δ_{step} are constants.

2. *The Ibots collectively carry out a trial with the newly generated programs* $New^i(t)s$.

At the beginning of the trial, the Ibots are positioned at a random configuration (which may be clustered or scattered). At the end of the trial, each Ibot returns $N^i_{in}(t)$. The team integral estimate is $\hat{I}(t) = N_{in}(t)/N_{max}$, with $N_{in} = \sum_i N^i_{in}$. The error in the estimate $E(t)$ is:

$$E(t) = \frac{|I - \hat{I}(t)|}{\max(I, 1 - I)}$$

3. *The team reinforcement signal $R(t)$ is computed and communicated to each Ibot.*

$R(t)$ is defined as the difference between the team errors in two successive trials:

$$R(t) = E(t - 1) - E(t) \tag{3}$$

$E(t - 1)$ can be thought of as a naïve predictor of $E(t)$.

4. *Each Ibot "i" independently computes a modification for its control program and updates it.*

The modification is:

$$\Delta\alpha^i_{prog}(t) = \epsilon \cdot R(t) \cdot (\alpha^i_{new}(t) - \alpha^i_{new}(t - 1)) \tag{4}$$

$$\Delta\delta^i_{prog}(t) = \epsilon \cdot R(t) \cdot (\delta^i_{new}(t) - \delta^i_{new}(t - 1)) \tag{5}$$

where ϵ, a real parameter, is the learning rate.
Given:

$$\alpha^i_{temp}(t+1) = \alpha^i_{prog}(t) + \Delta\alpha^i_{prog}(t)$$
$$\delta^i_{temp}(t+1) = \delta^i_{prog}(t) + \Delta\delta^i_{prog}(t)$$

the updated control programs is:

$$\alpha^i_{prog}(t+1) = \begin{cases} 0° & \text{if } \alpha^i_{temp}(t+1) < 0° \\ \alpha_{max} & \text{if } \alpha^i_{temp}(t+1) > \alpha_{max} \\ \alpha^i_{temp}(t+1) & \text{otherwise} \end{cases}$$

$$\delta^i_{prog}(t+1) = \begin{cases} 0 & \text{if } \delta^i_{temp}(t+1) < 0 \\ \delta_{max} & \text{if } \delta^i_{temp}(t+1) > \delta_{max} \\ \delta^i_{temp}(t+1) & \text{otherwise} \end{cases}$$

The description of the algorithm is completed by the following remarks and definitions.

- When the Ibots work with public control programs, only one Ibot generates the tentative program $New^i(t)$ at step 1; then, $New^i(t)$ is communicated to the other team members.
- About the error measure $E(t)$: by dividing $|I - \hat{I}(t)|$ by $max(I, 1 - I)$, $E(t)$ varies between 0 and 1, no matter what the value of I. As a consequence, the reinforcement signal also varies in a fixed interval, namely $[-1, +1]$.
- $E(-1) = E(0)$. At trial 0, we assume that the expected error $E(-1)$ is equal to the measured error $E(0)$. This implies $R(0) = 0$, so no change is made to $Prog(0)$.
- $\rho(t) = max(|R(t-1)|, \rho_c)$. The amount of variation in the new control programs is proportional to the absolute value of the reinforcement signal at previous trial. This is to enhance the tendency of escaping from programs with unpredictable performance, and, viceversa, to favor the convergence towards programs with stable performance. ρ_c is a positive constant which maintains a minimal level of exploration in program space when $R(t-1) = 0$.

We conclude this section by placing this experiment in the reinforcement learning panorama. We have set the stage for a *nonassociative, immediate* reward learning experiment. This experiment qualifies as nonassociative, because there is no perception-action mapping to be learned: the only input to the learners is the reinforcement signal. Moreover, since this is fed to the Ibots as soon as the trial is finished and the trial is the team's atomic action, this is immediate reinforcement learning. Equation 3 establishes that the payoff is positive if the $New^i(t)$s provided a better estimate of the integral than the $New^i(t-1)$s, negative if the estimate was worse, zero if it was equal (equally good or equally bad). Thus $R(t)$ decides the direction of change in the control programs. As an example[2], if $\alpha_{new}^i(t)$ is greater than $\alpha_{new}^i(t-1)$, $\alpha_{prog}^i(t+1)$ will increase or decrease with respect to $\alpha_{prog}^i(t)$ depending on whether $R(t)$ was positive or negative (Eq. 4). On the contrary, if $\alpha_{new}^i(t)$ is less than $\alpha_{new}^i(t-1)$, $\alpha_{prog}^i(t+1)$ will increase if $R(t)$ is negative, decrease if $R(t)$ is positive. In all cases the direction of change in the control programs is meant to increase the probability of those $Prog^i$s which proved to be better in the integral estimate process. Moreover, $Prog^i$s remain unchanged when $R = 0$. This learning method is a non-associative version of the basic, connectionist reinforcement learning algorithm proposed in [19]. Overall the learning algorithm is expected to guide the team towards admissible and stable control programs.

6 Experiments

On the "half-full" *Region* ($I = 0.49$) of Fig. 1(a), we have run repeated learning experiments with Ibots' teams of increasing size ($N_{ibots} = 1, \ldots, 14$), with public or private $Prog^i$s, and starting from clustered or scattered configurations. We have also considered the situation where the Ibots' skills are no longer homogeneous because of differences in sensing and acting capabilities. In all experiments we have set: $N_{max} = 100$ (the corresponding admissible error being 0.1), $\alpha_{max} = 180°$ and $\delta_{max} = 500$ (length of arena side), $\alpha_{step} = \alpha_{max}/10$ and $\delta_{step} = \delta_{max}/10$, $\rho_c = 0.1$, and, for each Ibot i, $Prog^i(0) = (\alpha_{prog}^i(0), \delta_{prog}^i(0)) = (0°, 0)$: this "staying in place" program was chosen *to bias* the Ibots towards "economic" programs, i.e. programs requiring small translations. To examine the form of the learnt programs, a learning experiment was stopped either after having obtained 10 consecutive admissible estimates, or after a fixed number of trials, depending on which of these two events occured first. The former stopping criterion was introduced for convenience, for not devoting too much time to experiments which did converge rapidly.

6.1 One Ibot

The single Ibot experiment is a point of reference for comparing results obtained with teams of Ibots. It requires to discover a pair $(\alpha_{prog}, \delta_{prog})$ which produces admissible and stable integral estimates. As the program space searched by the learning algorithm is bidimensional, one can explore it in a systematic way to test the quality of a significant number of programs. Thus, before starting the actual learning experiments, we have run background trials with all combinations of α_{prog} and δ_{prog}, with α_{prog} ranging in $\{0°, 10°, \ldots, 180°\}$, and δ_{prog} ranging in $\{0, 25, \ldots, 500\}$. Each combination program was tested on N_{trials} different trials to have a sample of integral estimates $\{\hat{I}_k, k = 1, \ldots, N_{trials}\}$. Then, for each program we computed the *mean of errors* $\mu_{prog}(E)$ and the *variance of errors* $\sigma_{prog}^2(E)$:

---

[2] All the following observations made on α_{new}^i hold for δ_{new}^i too, see Eq. 5.

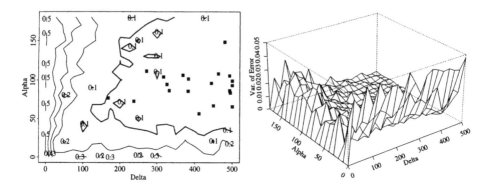

Fig. 2. One Ibot on the "half-full" *Region*. (Left) Contour plot of $\mu_{prog}(E)$ and convergence points for 20 learning experiments. (Right) Plot of $\sigma^2_{prog}(E)$.

$$\mu_{prog}(E) = \frac{1}{N_{trials}} \sum_k |I - \hat{I}_k| = \frac{1}{N_{trials}} \sum_k E_k$$

$$\sigma^2_{prog}(E) = \frac{1}{N_{trials}} \sum_k (E_k - \mu_{prog}(E))^2$$

$\mu_{prog}(E)$ and $\sigma^2_{prog}(E)$ describe the suitability of the corresponding program for the integration task: $\mu_{prog}(E)$ gives an indication on the accuracy of the estimates, while $\sigma^2_{prog}(E)$ measures their stability.

Figure 2 shows the plots in program space of $\mu_{prog}(E)$ (left) and of $\sigma^2_{prog}(E)$ (right): the program space axes are indexed by δ_{prog} and α_{prog}. On the $\mu_{prog}(E)$ plot we have highlighted the contour lines of level 0.1: these lines identify the space of admissible programs. Notice that these programs are also stable (see $\sigma^2_{prog}(E)$ plot). Most of them are distant from the "programmer solution" (180°, 700): admissible and stable solutions start at (50°, 200). Observe also that large values of α_{prog} require large values for δ_{prog} to be admissible, because small values of δ_{prog} would confine the Ibot's motion to a localized area (as an example, see the gray trace of Ibot 1 in Fig. 1 (c)). Finally, from the $\sigma^2_{prog}(E)$ plot we remark that not only admissible programs are stable. For example, all "staying in place" programs ($\delta_{prog} = 0$) have a very predictable performance. This makes the learning task more difficult as the Ibot's initial program $Prog(0) = (0°, 0)$ acts as a local minimum with respect to the stability criterion.

The results of 20 different learning experiments for the single Ibot have been overlaid on the left plot of Fig. 2. The convergence point of each experiment is indicated by a square. Learning stopped in all cases before the limit of 1000 trials. All experiments ended inside the space of admissible and stable programs.

For the sake of comparison, Fig. 3 shows the plots of $\mu_{prog}(E)$ (left) and of $\sigma^2_{prog}(E)$ (right) for one Ibot running on the "chessboard" *Region* ($I = 0.49$,

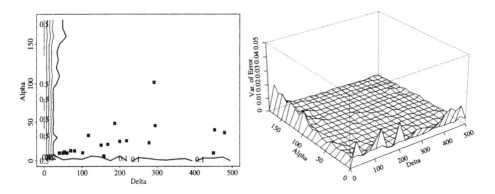

Fig. 3. One Ibot on the "chessboard" *Region*. (Left) Contour plot of $\mu_{prog}(E)$ and convergence points for 20 learning experiments. (Right) Plot of $\sigma^2_{prog}(E)$.

Fig. 1 (d)). The space of admissible programs is considerably larger than the one for the "half-full" *Region*, because the integral of the "chessboard" *Region* can be predicted even by a localized motion. Hence, the Ibot can afford travelling shorter distances to do a good job (Fig. 1 (d)).

6.2 Teams of Ibots

The form of admissible and stable programs completely changes for teams of Ibots.

Public Control Programs. The first case study we have addressed is that of Ibots equipped with public control programs. As in the single Ibot case, the program space is bidimensional, so we first ran background trials (with no learning) for teams of increasing size, both for the clustered and the scattered configuration. The results for the largest team of 14 Ibots are shown in Fig. 4. The left plot is the contour plot of $\mu_{prog}(E)$ for the clustered configuration, the right plot shows $\mu_{prog}(E)$ for the scattered configuration. On the left plot, the bold line delimits the space of admissible programs; for the scattered configuration, all programs result to be admissible (all contour lines are below level 0.1).

By comparing these results with those of Fig. 2 (left), one observes how the single Ibot's solution space "shrinks" or "expands" depending on whether the Ibots are started in the clustered or in the scattered way. Why?

First, consider the clustered configuration starting condition. As the Ibots begin a trial from the same position and with the same orientation, they have to disperse in the arena in order to explore it. Moreover, the number of samples they are allowed to take as individuals decreases as the team size increases, because the samples budget N_{max} remains the same whatever the team size. As a consequence, in a large team, each Ibot is granted fewer samples and fewer

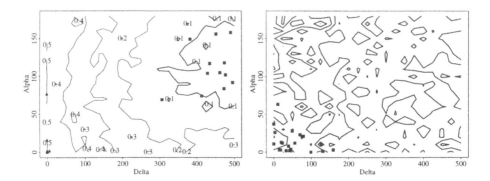

Fig. 4. A team of 14 Ibots with public control programs on the "half-full" *Region*. Contour plot of $\mu_{prog}(E)$ and convergence points for 20 learning experiments for clustered configurations (left) and for scattered configurations (right).

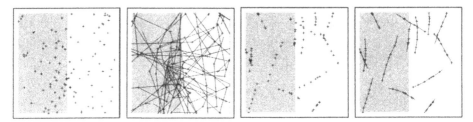

Fig. 5. A team of 14 Ibots running trials with learnt public control programs. (From left to right) (a) Samples and (b) trajectories for a team started in the clustered configuration and running the control program: $Prog = (100°, 450)$. (c) Samples and (d) trajectories for a team started in the scattered configuration and running the control program: $Prog = (10°, 50)$.

elementary movements to disperse in the arena. Given this constraint, the only way of achieving dispersion in few movements is through control programs with large variability both in translation and in rotation. *In conclusion, most of the solutions which are valid for the single Ibot would not work for this team.*

Second, consider the opposite case where the Ibots start a trial from scattered positions. As they are already uniformly distributed in the arena, any kind of motion program would lead to admissible estimates. In principle, the solution which would profit the most from this favorable start would be to perform a very localized motion around each Ibot's initial position. *We stress that this team solution would not be admissible for the single Ibot.*

Figure 4 also reports the convergence points of 20 learning experiments for both initial configuration types. Experiments which converged to admissible and stable programs within the time limit of 2000 trials are represented by squares,

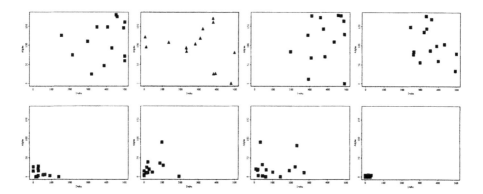

Fig. 6. A team of 14 Ibots with private control programs on the "half-full" *Region*. Convergence points for 4 learning experiments for clustered configurations (first row) and for scattered configurations (second row). For each experiment, the 14 private programs learnt by the team members are shown. In each plot, the vertical axis is indexed by α_{prog}, the horizontal axis by δ_{prog}.

while non-converging experiments are represented by triangles. Not surprisingly, all the experiments for the scattered configuration converged very rapidly (right). On the contrary, for the clustered configuration, not all experiments managed to converge to admissible programs within the predefined time limit (left). This is due to the fact that the Ibots initial program is $Prog^i(0) = (0°, 0)$, a bad but very stable program. As a matter of fact, the stability of this program becomes stronger as the number of Ibots grows. Therefore, it may take a considerable amount of time for the team to get away from this inconvenient initial program.

Finally, figure 5 shows trajectories and samples taken by a clustered ((a) and (b)) and a scattered team ((c) and (d)) of 14 Ibots running trials with learnt public programs.

Private Control Programs. Figure 6 shows a representative set of learning experiments performed with a team of 14 Ibots working with private control programs and started from clustered (plots on the first row) or scattered configurations (plots on the second row). Each plot shows the programs learnt by the team within the time limit of 20000 trials; programs which did not converge have been represented by triangles (second plot on the first row).

Essentialy, these solutions are similar to those obtained with public control programs. Within the same category of initial configuration, Ibots learn the same typology of programs: clustered Ibots need large variability in angle and translation, while scattered Ibots don't. This uniformity in the shape of the solutions is not surprising, because the integration task does not require differentiation in behavior as long as the Ibots have homogeneous skills.

From the point of view of learning, the main difference between dealing with a single public program or with many private programs is the robot credit assign-

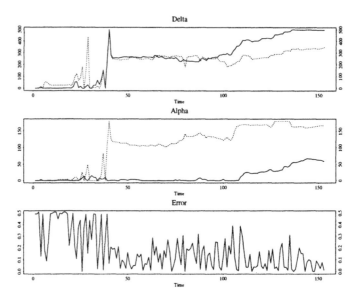

Fig. 7. Robot credit assignment problem for a team of two Ibots: Ibot 1 (dashed line) and Ibot 2 (solid line). Time evolution of δ^i_{prog}s (first plot), of α^i_{prog}s (second plot), and of the error E (third plot).

ment problem, which arises when the Ibots learn personal programs. Figure 7 illustrates the robot credit assignment problem for a team of two Ibots. The first and the second time plot present the evolution of the Ibots' δ^i_{prog}s and α^i_{prog}s parameters, respectively; the third time plot shows the error E in the integral estimate produced by the team. Observe that, around time 30, Ibot 1 (dashed line) has already acquired an admissible program ($Prog^1 = (49°, 437)$), but this does not appear at level of team performance because Ibot 2 (solid line) is still locked to the initial "staying in place" program. Consider also that Ibot 2's contribution to the team integral weighs more than Ibot 1's contribution, because Ibot 2 takes more samples: "bad Ibots count more". This is also the cause for the non-converging experiment of Fig. 6, where a minority of Ibots translating for short distances damage the team performance. To improve this state of affairs, Ibot 1 (Fig. 7), first backtracks from its admissible program, then relearns at a similar pace with Ibot 2.

Private Control Programs for Heterogeneous Ibots. As a last experiment, we wanted to make the learnt control programs more specialized. A way of achieving this is by differentiating the Ibots' individual skills. Figure 8 refers to a learning experiment with a team of two heterogeneous Ibots. Ibot 1 (dashed line) translates four times as fast as Ibot 2 (solid line). Moreover, Ibot 2 is *blind*: its ground color sensor reads "white" whatever its position in the arena.

The strategy discovered by the team to provide admissible and stable esti-

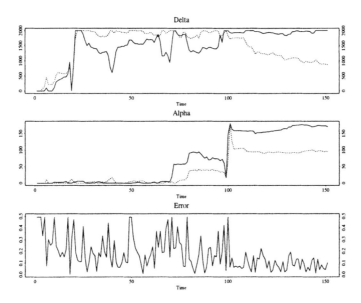

Fig. 8. A team of two heterogeneous Ibots: Ibot 1 (dashed line) moves 4 times faster than Ibot 2 (solid line), which is blind. Time evolution of δ^i_{prog}s (first plot), of α^i_{prog}s (second plot), and of the error E (third plot).

mates is clear from the δ^i_{prog}s and α^i_{prog}s plots of Fig. 8. The blind Ibot minimizes its catastrophic contribution to the team integral estimates by travelling long distances ($\delta^2_{prog} = 2000$, having set $\delta_{max} = 2000$ for this particular experiment). Observe that the error E stabilizes to low values only when the difference between δ^1_{prog} and δ^2_{prog} is sufficiently large. Still, Ibot 1 can afford a parameter of $\delta^2_{prog} = 800$ because it moves very fast.

Table 1 reports more programs learnt by this team in 5 repeated experiments. Ibot 1 always travels for shorter distances than Ibot 2. In all experiments, the balance between δ^1_{prog} and δ^2_{prog} is such that Ibot 1 consistently manages to collect at least 85 samples out of the 100 available to the team.

Finally, figure 9 shows the behavior of $\mu_{prog}(E)$ in program space of Ibot 1, when Ibot 2 works with a fixed control program. On the left plot, Ibot 2 is running $Prog^2 = (0°, 2000)$. This gives Ibot 1 the possibility of choosing its admissible program in a rather large set of programs: it can easily run up to $\delta^1_{prog} = 1000$ because its speed is four times the speed of the teammate. However, if the blind Ibot reduces its translations to $\delta^2_{prog} = 1000$ (right plot), the fast Ibot is forced to fit its control program to a reduced space of admissible programs.

7 Conclusions

The overall objective of the Ibots experiment was to understand how to use reinforcement learning to program automatically a team of robots with a common

δ^1_{prog}	α^1_{prog}	δ^2_{prog}	α^2_{prog}	$\mu(N^1_{sam})$	$\sigma(N^1_{sam})$
565	95	1528	179	85.5	1.5
876	95	2000	171	85.9	1.7
361	176	1549	2	88.9	1.3
537	165	1646	102	86.9	2.0
422	75	1061	66	84.1	0.5

Table 1. (Columns 1–4) Programs learnt by the heterogeneous team in 5 repeated experiments (one experiment per row). (Columns 5 and 6) Mean and standard deviation of the number of samples taken by Ibot 1 over 10 trials.

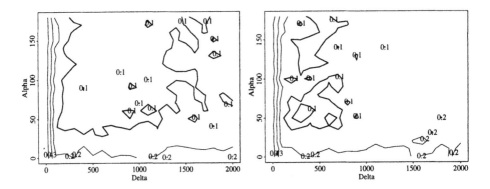

Fig. 9. Contour plots of $\mu_{prog}(E)$ in program space of Ibot 1 when Ibot's 2 program is: (Left) $Prog^2 = (0°, 2000)$ or (right) $Prog^2 = (0°, 1000)$.

mission. In addition, we wanted to derive *real team solutions*.

The "integration" mission of the Ibots is an artificial task for robots. However, the mission could also be interpreted as an exploration task, where the Ibots learn patterns of movement to reliably collect evidences about the region extension. Interestingly, it has been pointed to us [20] that the Ibots resemble networks of patrolling ants engaged in the task of monitoring events occurring throughout their territory [21].

The learning scenario for the Ibots is applicable to other missions because it relies on weak assumptions. A *team reinforcement signal* evaluates the behavior of the group as a whole; a single Ibot has no direct way of assessing its own performance, as distinct from the performance of its teammates. A *limited, common resource* constrains the Ibots, and there is no a priori rule to decide how this resource should be shared. When working with private control programs, the Ibots are *unaware* of teammate programs; during learning, each Ibot changes its own program independently, and has no information on how teammates are changing theirs.

As a general conclusion, experiments have demonstrated how different mission conditions require completely different control programs, and that a simple reinforcement learning procedure can find them. The key issue is: to optimize team performance instead of individual performance.

As far as the specific Ibots experiment is concerned, we cannot claim that the "general pattern" of the solutions discovered through learning were completely unexpected. However, as robot programmers, we have only a limited intuition for program parameters tailored to specific mission conditions (i.e. for a specific *Region*, for a given team size, for a specific team configuration, or for a particular set of robot skills). Sometimes, we are able to specify "a priori" programs which work for every possible mission condition (like those of Eq. 1); but, in certain mission contexts, these general solutions look unnatural. To write "ad hoc" programs for robots, a programmer will usually need to learn by trial-and-error himself: therefore, why not consider letting the robots do this, i.e learn by trial-and-error on their own [22]? *Second*: in general, a program which is admissible for a single Ibot is not ammissible for a team of Ibots, and viceversa. Thus, we cannot simply find a solution for one Ibot and clone it n times, n being the number of team members. The form of the solution to a problem changes as the number of "problem solvers" changes. Moreover, the robots become aware of this fact only if they are confronted with their performance as a team. On the contrary, a group of robots learning from individual payoffs would ignore opportunities which become evident only if the task is considered at a team level. *Third*, the space of admissible programs strongly depends on the number of Ibots involved in the mission and on their initial configuration in the arena. The admissibility space "shrinks" (and learning requires more time) when the Ibots are started in the clustered configuration and the team size grows. The increased difficulty is due to the fact that the number of samples is shared. On the contrary, the admissibility space "enlarges" (and learning requires less time) when the Ibots are started in the scattered configuration and the team size grows. The mission becomes easier in this case because, by initially distributing the Ibots at random in the arena, we bring them close to the problem solution; a team of individualistic robots would not be aware of this opportunity. *Fourth*, when the Ibots work with private control programs, the robot credit assignment problem arises, resulting in longer learning time. Interestingly, the robot credit assignment problem forces the Ibots to learn admissible programs at a similar pace, to prevent "slow" learners from jeopardizing the team mission. *Fifth*, the robot credit assignment problem vanishes when the Ibots learn a shared policy, and the learnt policy is still a real team solution. The possibility of learning a single public program instead of several private programs should be not overlooked in missions where specialization of the robot behavior is not required, because the time necessary for the team to learn a public program is much shorter. Finally, *sixth*, the Ibots with their heterogeneous acting and sensing capabilities manage to specialize their control programs so as to take advantage of their skills and to minimize the impact of their weaknesses.

Acknowledgements

Cristina Versino is supported by project No. 2129-042413.94/1 of the Fonds National de la Recherche Scientifique, Berne, Suisse.

References

1. A.G. Barto, R.S. Sutton, and C.J.C.H. Watkins. Learning and sequential decision making. Technical Report COINS-89-95, Dept. of Computer and Information Science, University of Massachusets, Amherst, 1989.
2. G. Weiß. Adaptation and learning in multi-agent systems: Some remarks and a bibliography. In G. Weiss and S. Sen, editors, *Adaptation and Learning in Multi-Agent Systems*, volume 1042, pages 1–21. Springer-Verlag, Lecture Notes in Artificial Intelligence, 1996.
3. M.J. Matarić, 1995. Personal Communication.
4. L.E. Parker. The effect of action recognition and robot awareness in cooperative robotic teams. In *IROS95, IEEE/RSJ International Conference on Intelligent Robots and Systems, Pittsburgh, PA, August*, volume 1, pages 212–219, 1995.
5. L.E. Parker. ALLIANCE: An architecture for fault tolerant, cooperative control of heterogeneous mobile robots. In *IROS94, IEEE/RSJ International Conference on Intelligent Robots and Systems, Munich, Germany, September*, pages 776–783, 1994.
6. M.J. Matarić, M. Nilsson, and K.T. Simsarian. Cooperative multi-robot box-pushing. In *IROS95, IEEE/RSJ International Conference on Intelligent Robots and Systems, Pittsburgh, PA, August*, 1995.
7. C.R. Kube and H. Zhang. Collective robotics: From social insects to robots. *Adaptive Behavior*, 2(2):189–218, 1994.
8. M.J. Matarić. Interaction and intelligent behavior. Technical Report AI-TR-1495, MIT Artificial Intelligence Lab, Boston, 1994.
9. M.J. Matarić. Learning in multi-robot systems. In G. Weiss and S. Sen, editors, *Adaptation and Learning in Multi-Agent Systems*, volume 1042, pages 152–163. Springer-Verlag, Lecture Notes in Artificial Intelligence, 1996.
10. M.J. Matarić. Learning to behave socially. In Meyer J.A. and S. Wilson, editors, *From Animals to Animats: International Conference on Simulation of Adaptive Behavior*, pages 453–462, Cambridge, MA, 1994. MIT Press.
11. A. Murciano and J. del R. Millán. Learning signaling behaviors and specialization in cooperative agents. *Adaptive Behavior*, 5(1):5–28, 1997.
12. R.H. Crites and A.G. Barto. Improving elevator performance using reinforcement learning. In D.S. Touretzky, M.C. Mozer, and M.E. Hasselmo, editors, *Advances in Neural Information Processing Systems 8*, pages 1017–1023, Cambridge MA, 1996. MIT Press.
13. J. Schmidhuber. A general method for incremental self-improvement and multi-agent learning in unrestricted environments. In X. Yao, editor, *Evolutionary Computation: Theory and Applications*. Scientific Publ. Co., Singapore, 1996.
14. C. Versino and L.M. Gambardella. Ibots: Learning real team solutions. In *IC-MAS96 Workshop on Learning, Interaction and Organizations in Multiagent Environments*, Kyoto, Japan, December 1996.

15. L.E. Parker. L-ALLIANCE: A mechanism for adaptive action selection in heterogeneous multi-robot teams. Technical Report ORNL/TM-13000, Oak Ridge National Laboratory, Tennessee, November 1995.

16. J. C. Deneubourg, S. Goss, N. Franks, A. Sendova-Franks, C. Detrain, and L. Chrétien. The dynamics of collective sorting: Robot-like ants and ant-like robots. In Meyer J.A. and S. Wilson, editors, *From Animals to Animats: International Conference on Simulation of Adaptive Behavior*, pages 356–363. MIT Press, 1991.

17. J.M. Hammersley and D.C. Handscomb. *Monte Carlo Methods*. Methnen & Co., London, 1964.

18. A. Tucker. *Applied Combinatorics*. John Wiley & Sons, New York, 1995.

19. A.G. Barto, R.S. Sutton, and P.S. Brouwer. Associative search network: A reinforcement learning associative memory. *Biological Cybernetics*, 40:201–211, 1981.

20. A. Martinoli, 1996. Personal Communication.

21. F.R. Adler and D.M. Gordon. Information collection and spread by networks of patrolling ants. *The American Naturalist*, 140(3):373–400, 1992.

22. L.P. Kaelbling. *Learning in Embedded Systems*. MIT Press, Cambridge, MA, 1993.

Learning by Linear Anticipation in Multi-Agent Systems

Paul Davidsson

Department of Computer Science, University of Karlskrona/Ronneby
S–372 25 Ronneby, Sweden
paul.davidsson@ide.hk-r.se   http://www.ide.hk-r.se/~pdv

Abstract. A linearly anticipatory agent architecture for learning in multi-agent systems is presented. It integrates low-level reaction with high-level deliberation by embedding an ordinary reactive system based on situation-action rules, called the Reactor, in an anticipatory agent forming a layered hybrid architecture. By treating all agents in the domain (itself included) as being reactive, this approach reduces the amount of search needed while at the same time requiring only a small amount of heuristic domain knowledge. Instead it relies on a linear anticipation mechanism, carried out by the Anticipator, to learn new reactive behaviors. The Anticipator uses a world model (in which all agents are represented only by their Reactor) to make a sequence of one-step predictions. After each step it checks whether an undesired state has been reached. If this is the case it will adapt the actual Reactor in order to avoid this state in the future. Results from simulations on learning reactive rules for cooperation and coordination of teams of agents indicate that the behavior of this type of agents is superior to that of the corresponding reactive agents. Also some promising results from simulations of competing self-interested agents are presented.

1   Introduction

Some experimental results on multi-agent learning using a linearly anticipatory agent architecture are presented. This architecture, thoroughly described in Davidsson [2], is an instantiation of the general framework of anticipatory agents suggested by Astor, Davidsson, and Ekdahl [3, 6].

1.1   Anticipatory Agents

The framework for anticipatory agents is based on the concept of anticipatory systems as described by Rosen [13]. It is a hybrid approach that synthesizes low-level reactive behavior and high-level symbolic reasoning. As illustrated in Figure 1, an anticipatory agent consists of three main entities: a reactive system, a world model, and a meta-level component. The world model should, in addition to the description of the agent's environment, also include a description of the reactive part of the agent and of all other agents in the environment. The basic idea is that the meta-level component, which we will refer to as the *Anticipator*, makes use of the world model to make a sequence of predictions of future states. These predictions are then used by the Anticipator to guide the agent's behavior on a high-level by adapting the reactive component, the *Reactor*, which controls the low-level behavior. Thus, the working of an anticipatory agent can

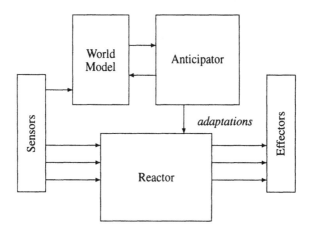

Fig. 1. The basic architecture of an anticipatory agent.

be viewed as two concurrent processes, one reactive at the object-level and one more deliberative at the meta-level.

The framework for anticipatory agents do not specify exactly how the predictions should be used to modify the Reactor's properties. Below we will present A Linearly Quasi-Anticipatory Agent Architecture (ALQAAA) that adopts the following approach, suggested by Rosen as being the most basic: The state space is partitioned into regions corresponding to "desirable" and "undesirable" states. As long as the predicted states of the World Model remains in a desirable region, no action is taken by the Anticipator. As soon as the sequence of predicted states moves into an undesirable region, the Anticipator is activated to change the dynamics of the Reactor in such a way as to keep the agent out of the undesirable region.

Rosen argues that a system is anticipatory in the strict sense if its world model is a perfect description of the environment (including all agents). However, as this in general is not possible, he also introduces systems with imperfect world models and calls the behavior of such systems quasi-anticipatory.

2 Linearly Anticipatory Agents

In linearly anticipatory agents, the Reactor and the Anticipator are run (asynchronously) as two separate processes. The Reactor process is given a high priority whereas the Anticipator is a low priority process that runs whenever the Reactor is "waiting", e.g., for an action to be performed. Since the Reactor is able to preempt the Anticipator at any time, reactivity is always guaranteed. Thus, the Anticipator has to be a kind of *anytime algorithm* [4], or rather anytime process, in that it should always be able to return a result when it is interrupted.[1] The appropriateness of using anytime algorithms in au-

---

[1] According to Dean and Boddy [4], the main characteristics of anytime algorithms are that "... (i) they lend themselves to preemptive scheduling techniques (i.e., they can be suspended and

tonomous agent contexts where real-time requirements are common has been pointed out by, for example, Zilberstein and Russell [15] and Bresina and Drummond [5].

The Reactor carries out a never ending cycle of: perception of the environment, action selection by situation-action rules, and performance of action. The basic algorithm of the Reactor is given below:

process REACTOR;
while true **do**
    Percepts ← Percieve;
    Action ← SelectAction(Percepts);
    Perform(Action);

The Anticipator, on the other hand, carries out a never ending cycle of anticipation sequences. Each such sequence begins with making a copy of the World Model, including descriptions of all agents as physical entities in the environment and their sets of reaction rules. These descriptions are then used to make a sequence of one-step predictions. After each prediction step, it is checked whether an undesired state has been reached, or whether the agent has achieved its goal. If the simulated agent has reached an undesired state, the actual Reactor will be adapted in order to avoid reaching this state. The basic algorithm of the Anticipator is as follows:

process ANTICIPATOR;
while true **do**
    WorldModelCopy ← WorldModel;
    UndesiredState ← false;
    while not UndesiredState **and not** GoalAchieved(WorldModelCopy) **do**
        for each Agent **do**
            Percepts ← WorldModelCopy(Agent).Percieve;
            Action ← WorldModelCopy(Agent).SelectAction(Percepts);
            WorldModelCopy(Agent).Perform(Action);
        UndesiredState ← Evaluate(WorldModelCopy);
    if UndesiredState **then**
        Adapt(Reactor);

Note that since the behavior of the Reactor in each situation is determined by situation-action rules, the Anticipator always assumes that it knows which action the Reactor will perform. To achieve this, also the environment and all other agents are treated as being purely reactive. (Some of the consequences of this simplification are discussed in Section 5.) Thus, since everything is governed by situation-action rules, the anticipation

resumed with negligible overhead), (ii) they can be terminated at any time and will return some answer, and (iii) the answers returned improve in some well-behaved manner as a function of time." (p.52) mechanism requires no search, or in other words, the anticipation is *linear*. It should also be noted that the goal of the agent is not limited to have only a singular goal. In a multi-goal scenario, some of the changes (modifications) of the Reactor should only hold for a limited interval of time, e.g., until the current goal has been achieved. Otherwise, there is a danger that these changes might prevent the agent to achieve other goals.

A linearly anticipatory agent can be specified as a tuple, $\langle \mathcal{R}, \mathcal{W}, \mathcal{U}, \mathcal{M} \rangle$, where:

$\mathcal{R}$ is the set of situation-action rules defining the Reactor.

$\mathcal{W}$ is the description of the environment (including all agents).

$\mathcal{U}$ is the set of undesired states.

$\mathcal{M}$ is the set of adaptation rules describing how to modify $\mathcal{R}$.

The Anticipator is defined by $\mathcal{U}$ and $\mathcal{M}$. For each element in $\mathcal{U}$ there should be a corresponding rule in $\mathcal{M}$, which should be applied when an undesired state of this kind is anticipated. Thus, we need in fact also a function, $f : U \rightarrow M$, that determines which rule for modifying the Reactor that should be applied given a particular type of undesired state. However, as this function typically is obvious from the specification of $\mathcal{U}$ and $\mathcal{M}$, it will not be described explicitly. Moreover, in all simulations described below, $\mathcal{W}$ will consist of the positions of all obstacles, targets, and agents present in the environment together with Reactor descriptions of all agents. Using these terms, the function Evaluate can be described as checking whether the current anticipated state belongs to $\mathcal{U}$, and Manipulate as first applying f on the anticipated undesired state and then using the resulting rule from $\mathcal{M}$ to modify $\mathcal{R}$.

3 Experiments

The problem domain has deliberately been made as simple as possible in order to make the principles of linearly anticipatory behavior as explicit as possible. The environment is a two-dimensional grid (10×10) in which a number of unit-sized square obstacles forms a maze. In addition to these static objects, there are two kinds of dynamic objects: agents, which can move about in the maze, and targets, which can be removed by an agent. The goal of an agent is to pick up the targets. To be able to pick up a target, an agent must be in the same position as the target. The agents are able to move in four directions (north, south, east, and west), unless there is an obstacle that blocks the way. An agent is always able to perceive the direction to a target, and whether there are any obstacles (or agents) immediately north, south, east, or west of the agent. Similar environments have been used in single-agent experiments by many researchers (cf. Kinny and Georgeff [10] and Sutton [14]). (The advantages and disadvantages of this kind of testbeds have been discussed at length by Hanks, Pollack and Cohen [9].) The generalization of the environment into a multi-agent scenario was suggested by Zlotkin and Rosenschein [16].

Simulations of ALQAAA agents in single-agent domains are described in Davidsson [2]. It was concluded that (if the Anticipator is given enough time to anticipate) the following simple agent specification suffice to achieve optimal behavior in such domains:

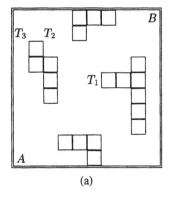 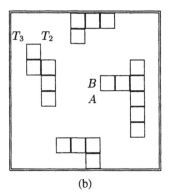

(a)   (b)

Fig. 2. The behavior of two competing reactive agents: (a) the initial state (b) the situation after 8 time steps, agent B has picked up target T_1.

$\mathcal{R} = \{move\ to\ the\ free\ position\ closest\ to\ the\ nearest\ target,\ but\ do\ not\ turn\ around\ 180°$ *unless forced to*$\}, \mathcal{U} = \{being\ in\ a\ loop\}$, and $\mathcal{M} = \{avoid\ the\ position\ in\ the\ loop\ closest\ to\ the\ target\}$. All agents in the multi-agent experiments described in the following two sections are based on this specification. We will point out the advantages of being anticipatory, both when competing and when cooperating with other agents. Although the experiments have been carried out with only two (competing or cooperating) agents in order to make things as clear as possible, it would be trivial to extend the experiments to permit a larger number of agents.

3.1 Competing Agents

The main idea is that a self-interested ALQAAA agent should use its knowledge about the behavior of other agents in order to detect future situations in which other agents interfere with the agent's own intentions (i.e., goals). If such a situation is detected, the Anticipator should adapt the Reactor in order to minimize the probability that this situation will ever occur.

The goal of both the agents (where A is "our" agent and B its opponent) is to pick up as many targets as possible. A good strategy for an ALQAAA agent would be that, when the Anticipator realizes that B will reach a target before A, it adapts the Reactor so that it will ignore this target. Thus, we have that: $\mathcal{U} = \{being\ in\ a\ loop,\ pursuing\ targets\ that$ *presumably will be picked up by another agent*$\}$ and $\mathcal{M} = \{avoid\ the\ position\ in\ the$ *loop closest to the target, avoid the target that presumably will be picked up by another agent*$\}$. For example, if the target to avoid is named T_i, $\mathcal{R}$ will be modified into: $\{move$ *to the free position closest to the nearest target except for* T_i, *but do not turn around* $180°\ unless\ forced\ to\}$.

However, let us first see how two reactive agents, which both correspond to $\mathcal{R}$ as defined above will behave. An environment containing three targets is described in Figure 2(a). If the agents start at the same time the following will happen. Both agents perceive that T_1 is their closest target and head towards it. As B is somewhat closer to

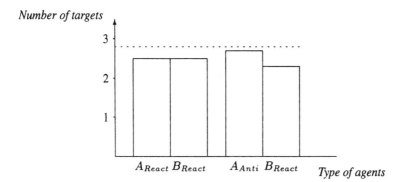

Fig. 3. Comparison between two sets of competing agents. To the left are both A and B reactive agents and to the right is A an ALQAAA agent and B a reactive agent. The vertical axis indicates the number of targets picked up by an agent (averages over 1000 runs). The estimated optimal number of targets that A is able to pick up in this situation (i.e., given B's behavior) is illustrated by the dashed line.

T_1 than A, it will reach it first and pick it up (see Figure 2(b)). B will then head for T_2 which now is the closest target. A will also head for T_2 following B. B then reaches T_2, picks it up, and heads for the last target T_3 with A still behind. Eventually B will pick up also T_3. Thus, B gets all the targets and A gets none.

If we, on the other hand, let A be an ALQAAA agent and start in the same position as above, A's Anticipator will soon detect that B will be the first to reach T_1. The Anticipator will then adapt the Reactor with the result that A will avoid this target and instead head towards T_3 (which is the next closest target to A). It will reach T_3 at the same time as B reaches T_1. When the agents have picked up their targets, there is only one target left (T_2). Since A is closest to T_2, A will reach it first. Thus, by anticipating the behavior of both itself and the other agent, A will in this case pick up two targets whereas B only picks up one.

There are also some quantitative results on the superiority of ALQAAA agents when competing with reactive agents. In the experiments there were 30 obstacles, 5 targets and two agents. In the first session both the agents were reactive and in the second there was one ALQAAA agent competing with a reactive one. The results shown in Figure 3 tell us that the performance indeed is improved (being almost optimal[2]) when the agent behaves in an anticipatory fashion.

3.2 Cooperating Agents

We shall now see how ALQAAA agents can be used for cooperation. The task for the two agents is to pick up all the targets in shortest possible time. It does not matter which agent that picks up a particular target. To begin with, let us apply the agents in the last example (i.e., A is an ALQAAA agent and B a reactive agent) to the situation described in Figure 4(a). As these agents are not cooperating, their global behavior will

---

[2] The optimal performance is estimated from an analysis of a subset of the experiment situations.

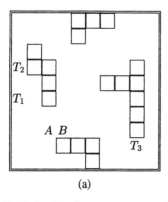 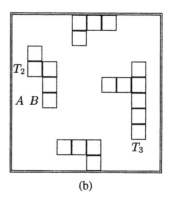

(a)   (b)

Fig. 4. The behavior of two non-cooperating agents one ALQAAA (A) and one reactive (B). (a) the initial state (b) the situation after 4 time steps, agent A has picked up target T_1.

(as one might expect) not be optimal. What will happen is that both agents initially head towards the same target (T_1). When agent A reaches T_1 we have the situation depicted in Figure 4(b). The other targets will then be approached in the same fashion, with one agent following the other. As a result, it will take these non-cooperating agents 15 time-steps to pick up all the targets.

Cooperating agents, on the other hand, should be able to make use of the fact that the ALQAAA agent knows that it will pick up the two closest targets. One way of doing this is to let agent A send a message to agent B (which still is a reactive agent) when it believes that it will pick up a particular target. This message contains the information that agent B should avoid this target. Thus, we add "*other agent pursuing target that presumably will be picked up by me*" to U and "*send message to other agent that it should avoid the target*" to M. The modifications of the reactive rules will be very similar to those in the last section. In this case, however, agent A adapts the behavior of agent B.

When this method is applied to the previous example, agent A will detect that it will pick up targets T_1 and T_2 and sends therefore messages to B that these should be avoided. B then directly heads towards T_3, which is the only remaining target that A will not reach before B. As a result, this system of cooperating agents will use only 6 time-steps to pick up all the targets. The quantitative results (using 30 obstacles, 5 targets and two agents) are summarized in Figure 5. We see that when the two agents are cooperating they come close to optimal behavior.

In the scenario described here, only A is an ALQAAA agent whereas B is an ordinary reactive agent. However, even if we let B be an ALQAAA agent defined in the same way as A, the performance would not be increased. The reason for this is that both agents would have made the same predictions and would therefore send messages to each other about things that they both have concluded by themselves. Still, in a more realistic setting where the agents do not have exactly the same information about the world, this approach would be more fruitful. This would correspond to *lateral co-ordination* in contrast to the *hierarchical co-ordination* used above. When using

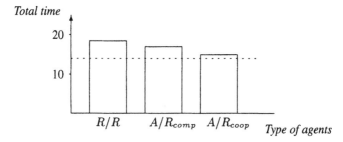

Fig. 5. Comparison between three sets of agents in terms of the total time it takes to collect all the five targets (averages over 1000 runs). R/R denotes two reactive agents, A/R_{comp}, one reactive and one ALQAAA agent competing, and A/R_{coop}, one reactive and one ALQAAA agent that are cooperating. The estimated optimal time for two agents to collect all targets in this situation is illustrated by the dashed line.

lateral co-ordination things get quite complicated if one ALQAAA agent simulates the anticipatory behavior of another ALQAAA agent which in turn simulates the first agent's anticipatory behavior. The solution I suggest is to simulate only the reactive component of other agents and when an agent adapts its reactive component, it should communicate (e.g., broadcast) information about this adaptation to the other agents. In this way we are still able to make linear anticipations. This approach can be contrasted with the Recursive Modeling Method suggested by Gmytrasiewicz and Durfee [8] in which an agent modeling another agent includes that agent's models of other agents and so on, resulting in a recursive nesting of models.

4   Related Work

The main task of the Anticipator is to avoid undesired states whereas the main task of the Reactor is to reach the desired state(s). In other words, the Anticipator's goals are goals of maintenance and prevention rather than of achievement. Compare this to Minsky's suppressor-agents, discussed within his Society of Mind framework [11], which waits until a "bad idea" is suggested and then prevents the execution of the corresponding action. However, there is a big difference, suppressor-agents are not predictive. The Anticipator takes actions beforehand so that the bad idea never will be suggested! Thus, an Anticipator can be regarded as predictive suppressor-agent.

In conformity with the Sequencer component in Gat's ATLANTIS architecture [7], the Anticipator can be viewed as being based on the notion of *cognizant failures* (i.e., a failure that can be detected by the agent itself). However, an Anticipator detects these failures in a simulated reality (i.e., a model of the world), whereas a Sequencer has to deal with real failures.

The notions of Reactor and Anticipator have some similarities with the Reactor and Projector components in the ERE architecture suggested by Bresina and Drummond [5]. In particular, the Reactor in ERE is able to produce reactive behavior in the environment independently, but also takes advise from the Projector based on the Projector's explorations of possible futures. However, the Projector is more similar to a traditional planner in that it is based on search through a space of possible Reactor actions (a third component, the Reductor, is introduced to constrain this search), whereas the Anticipator simulates the behavior of the Reactor in its environment linearly (i.e., without search). Moreover, the Anticipator's main task is to avoid undesired states, whereas the Projector in the ERE tries to achieve desired states.

5 Conclusions

We have shown the viability of an approach for designing autonomous agents based on the concept of anticipatory systems called ALQAAA for simple multi-agent tasks. Learning is performed by letting the Anticipator component of the agent modify the Reactor component according to anticipated future states. The empirical results presented are promising but indicate that the behavior of the ALQAAA agents implemented has not been optimal. However, I have deliberately chosen very simple Reactors and Anticipators for the purpose of illustrating how easily the performance can be improved by embedding a Reactor in an ALQAAA agent. It should be clear that it is possible to develop more elaborate $\mathcal{R}$, $\mathcal{U}$, and $\mathcal{M}$ components that produce behavior closer to the optimal. In fact, it may even be possible to define a Reactor that produces optimal behavior (making $\mathcal{U}$ and $\mathcal{M}$ obsolete). However, such a solution would be inferior for at least two reason: (i) even if such an $\mathcal{R}$ do exist, it would probably be much more complex than the corresponding ALQAAA solution[3] (even in these very simple scenarios), and (ii) from a software engineering perspective the layered ALQAAA approach is preferable, since a clear distinction is made between implementing goal achieving behavior (specified by $\mathcal{R}$) and preventive behavior (specified by $\mathcal{U}$ and $\mathcal{M}$).

Compared to traditional planning, linear anticipation is a more passive way of reasoning. An ALQAAA agent just tries to predict what will happen if nothing unexpected occurs, whereas a planning agent actively evaluates the results of several possible actions. Consequently, planning agents relies heavily on search, whereas ALQAAA agents do not. The main reason for this is that all agents in the environment (also the ALQAAA agent itself) are treated as being reactive. In addition, it is interesting to note the small amount of heuristic domain knowledge given to the Reactor and the Anticipator (i.e., $\mathcal{R}$, $\mathcal{U}$, and $\mathcal{M}$). Thus, this approach drastically reduces the amount of search needed while at the same time requiring only a small amount of heuristic knowledge. Instead, it relies on a linear anticipation mechanism to learn complex behaviors.

In more realistic scenarios than the one described above, the anticipation will probably fail now and then. There are two possible reasons for such failures: either the world model, $\mathcal{W}$, is faulty, or the Anticipator, $\mathcal{U}$ or $\mathcal{M}$, is faulty. Failures of the first kind can be detected by comparing the world state at time $t+\Delta t$ to the predicted world state Δt time-steps ahead at time t. There are two ways of dealing with faulty world models: try to update $\mathcal{W}$, and if this is not possible, adapt (i.e., shorten) the anticipation length to keep the discrepancy between predicted and actual future states sufficiently low to make reliable predictions. Techniques useful for dealing with the problem of faulty models of

---

[3] One reason being that it is very difficult to detect undesired situations beforehand using only situation-action rules (especially those caused by other agents).

other agents have been analyzed in [1, 12]. A faulty Anticipator, on the other hand, can only be detected by noting that the agent does not achieve its goal. To do this we must introduce a higher level behavioral component that monitors the behavior of the agent. The simple solution is to let a human operator do the monitoring and redesign $\mathcal{U}$ or/and $\mathcal{M}$ when necessary. A more sophisticated solution would be to let a new component adapt the Anticipator in a similar way as the Anticipator adapts the Reactor.

The problems that the ALQAAA agents solved above can certainly be solved by other methods, but the point to be made is that we can qualitatively enhance the abilities of a reactive agent by embedding it in an ALQAAA agent. However, there are several obvious limitations to the application presented in this paper: (i) the environment is quite static (the only events that take place not caused by the agent itself are those caused by other agents), (ii) the agents have perfect models of the world and of other agents, and (iii) the agents have perfect sensors and the outcome of an action is deterministic.

Future work includes evaluation of the approach in other applications to see in which types of applications it performs well and whether there are any in which it is not appropriate.

References

1. D. Carmel and S. Markovitch. Opponent modeling in multi-agent systems. In G. Weiss and S. Sen, editors, *Adaptation and Learning in Multi-Agent Systems (LNAI 1042)*, pages 40–52. Springer Verlag, 1996.
2. P. Davidsson. *Autonomous Agents and the Concept of Concepts*. PhD thesis, Department of Computer Science, Lund University, Sweden, 1996.
3. P. Davidsson, E. Astor, and B. Ekdahl. A framework for autonomous agents based on the concept of anticipatory systems. In *Cybernetics and Systems '94*, pages 1427–1434. World Scientific, 1994.
4. T. Dean and M. Boddy. An analysis of time-dependent planning. In *AAAI-88*, pages 49–54. Morgan Kaufmann, 1988.
5. M. Drummond and J. Bresina. Anytime synthetic projection: Maximizing the probability of goal satisfaction. In *AAAI-90*, pages 138–144. MIT Press, 1990.
6. B. Ekdahl, E. Astor, and P. Davidsson. Towards anticipatory agents. In M. Wooldridge and N.R. Jennings, editors, *Intelligent Agents — Theories, Architectures, and Languages (LNAI 890)*, pages 191–202. Springer Verlag, 1995.
7. E. Gat. Integrating planning and reacting in a heterogeneous asynchronous architecture for controlling real-world mobile robots. In *AAAI-92*, pages 809–815. MIT Press, 1992.
8. P.J. Gmytrasiewicz and E.H. Durfee. Rational interaction in multiagent environment: Coordination. *(submitted for publication)*, 1996.
9. S. Hanks, M. Pollack, and P. Cohen. Benchmarks, testbeds, controlled experimentation, and the design of agent architectures. *AI Magazine*, 14(4):17–42, 1993.
10. D.N. Kinny and M.P. Georgeff. Commitment and effectiveness of situated agents. In *IJCAI-91*, pages 82–88. Morgan Kaufmann, 1991.
11. M. Minsky. *The Society of Mind*. Simon and Schuster, 1986.
12. Y. Mor, C.V. Goldman, and J.S. Rosenschein. Learn your opponent's strategy (in polynomial time!). In G. Weiss and S. Sen, editors, *Adaptation and Learning in Multi-Agent Systems (LNAI 1042)*, pages 164–176, 1996.
13. R. Rosen. *Anticipatory Systems – Philosophical, Mathematical and Methodological Foundations*. Pergamon Press, 1985.

14. R.S. Sutton. First results with Dyna, an integrated architecture for learning, planning and reacting. In W.T. Miller, R.S. Sutton, and P.J. Werbos, editors, *Neural Networks for Control*, pages 179–189. MIT Press, 1990.

15. S. Zilberstein and S.J. Russell. Anytime sensing, planning and action: A practical model for robot control. In *IJCAI-93*, pages 1402–1407. Morgan Kaufmann, 1993.

16. G. Zlotkin and J.S. Rosenschein. Coalition, cryptography, and stability: Mechanisms for coalition formation in task oriented domain. In *AAAI-94*, pages 432–437. MIT Press, 1994.

Learning Coordinated Behavior in
a Continuous Environment

Norihiko Ono and Yoshihiro Fukuta

Department of Information Science and Intelligent Systems
Faculty of Engineering, University of Tokushima
2-1 Minami-Josanjima, Tokushima 770, Japan

Abstract. Interesting efforts have been made to let multiple agents learn to appropriately interact, using various reinforcement-learning algorithms. In most of these cases, however, the state space for each agent is supposed discrete. It is not clear how effectively multiple reinforcement-learning agents are able to acquire appropriate coordinated behavior in continuous state spaces. The objective of this research is to explore the potential applicability of Q-learning in multi-agent continuous environments, when applied in conjunction with a generalization technique based on CMAC. We consider a modified version of the multi-agent block pushing problem, where two learning agents are interacting in a continuous environment to accomplish their common goal. To allow our agent to treat two-dimensional vector-valued inputs, we applied a CMAC-based Q-learning algorithm. This is a variant of L.-J.Lin's *QCON* algorithm. The objective is to incrementally elaborate a set of CMACs which can approximately provide the action value function under an optimal policy for the learning agent. The performance of our block pushing CMAC-based Q-learning agents is evaluated quantitatively and qualitatively through simulation runs. Although it is not intended to model any particular real world problem, the results are encouraging.

1 Introduction

Recently, various attempts have investigated the applicability of reinforcement learning to multi-agent problem domains(e.g., [2, 4, 5, 6, 7, 8, 9, 10, 11]). In most of these attempts, the environment for learning agents is supposed to have a discrete state space. The potentials and limitations of reinforcement-learning have not been clarified in multi-agent problem domains where the state space for each agent is continuous. In a continuous state space, learning agents are not able to explore the space exhaustively, and accordingly they are required to solve the *generalization* problem (often called the *structural* credit assignment problem) as well as the *temporal* credit assignment problem. The objective of this research is to explore the potentials of Q-learning[12] in multi-agent continuous environments, when applied in conjunction with a generalization technique based on *CMAC – Cerebellar Model Articulation Computer*[1].

2 Problem Domain

To investigate the potential applicability of Q-learning in multi-agent continuous environments, we consider a modified version of the *multi-agent block pushing problem*[8]. In this problem, two agents R_1 and R_2 independently operate attempting to move a block, b, from a starting position, S, to some goal area, G, following an optimal path P in Euclidean space as shown in Fig. 1. The two agents are engaged in the joint task, but they have no knowledge about the capabilities of each other. The agents have no knowledge about the system dynamics, but they can perceive their current positions in the space.

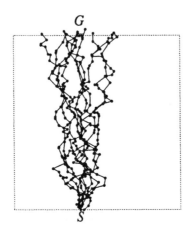

Fig. 1. Environment for learning agents and their learning trials.

At each time step, agent R_i is able to apply a force $\mathbf{F_i}(0 \leq |\mathbf{F_i}| \leq F_{max})$ to the block at an angle $\theta_i(0 \leq \theta_i \leq \pi)$. The resultant force is given by $\mathbf{F} = \mathbf{F_1} + \mathbf{F_2}$. We calculate the new position of the block by assuming a unit displacement per unit force along the direction of the resultant force.

The block pushing problem considered here is different from the original one[8] as follows: (i) our learning agents operate in a continuous environment, while agents in [8] operate in a discrete one, and (ii) our agents perceive their current position and receive a meaningful(non-zero) reinforcement signal only when getting out of the field, while agents in [8] perceive the distance between the block and the desired path and correspondingly receive an immediate reinforcement at each time step.

3 Learning Algorithm

In this problem domain, the state of an agent is a point in the two-dimensional Euclidean space. To allow a Q-learning agent to treat two-dimensional vector-valued inputs, we applied a *CMAC*-based Q-learning algorithm, which is a variant of L.-J.Lin's *QCON* algorithm[3]. The objective is to incrementally elaborate a set of CMACs that can approximately provide the action value function $Q_f(x, a)$, where f is an optimal policy for the learning agent. We suppose that the set of possible actions for an agent is discrete and its environment is deterministic. For each action a, we assign an independent CMAC, which models $Q_f(x, a)$ for any two-dimensional vector-valued input x. Let $CMAC(a)$ denote the CMAC associated with the action a.

We employ a simplified version of CMAC as shown in Fig. 2. Our CMAC consists of N *tilings*. A tiling is an $n \times n$ array of non-overlapping rectangular regions, called *tiles*, which cover the two-dimensional state space.

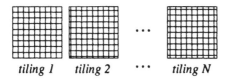

(a) A CMAC associated with an action a

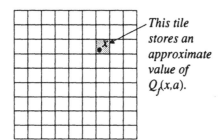

(b) Tiling: An array of non-overlapping Tiles

Fig. 2. A CMAC structure associated with a possible action a.

As shown in Fig. 3, the set of tilings constituting a CMAC is chosen such that any point in the space is contained in exactly N tiles, one from each tiling. We suppose each tile has an independent storage for holding an approximate value of action-value function $Q_f(x, a)$ where a is the action corresponding to the CMAC. Let the values held by the N tiles be

$$v_1, v_2, v_3, \ldots, v_N$$

then the value of $Q_f(x, a)$ is eventually estimated by an average over the N approximate values.

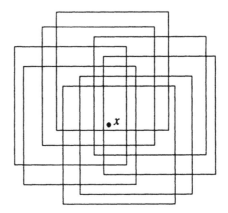

Fig. 3. Approximation of action value function by CMAC. Any point x is contained in just N tiles. The approximate value of the corresponding action value function $Q_f(x, a)$ is calculated by simply averaging those values stored in the tiles.

Given an input x, an agent selects and performs an action a, and correspondingly receives a reinforcement r and a new input y from the environment. Then, its learning task is to adjust the $\mathcal{CMAC}(a)$ so that its output approaches the exact value of $Q_f(x, a)$. Ideally, the following equation holds:

$$CMAC_a(x) = r + \gamma \max_{b \in A(y)} CMAC_b(y)$$

where $CMAC_a(x)$ denotes the action value function $Q_f(x, a)$ provided by the $\mathcal{CMAC}(a)$, $A(y)$ is a set of possible actions for the input y, and $\gamma(0 \le \gamma < 1)$ is the discount factor. The learning proceeds by slightly adjusting the $\mathcal{CMAC}(a)$ in order to reduce the difference between the two sides of the equation.

The $\mathcal{CMAC}(a)$ is adjusted as follows. Let

$$v_1, v_2, v_3, \dots, v_N$$

be the approximate values of $Q_f(x, a)$ being held in the tiles of the $\mathcal{CMAC}(a)$ which cover the point x. Then, the N values are respectively changed to

$$v_1 - \alpha e, v_2 - \alpha e, v_3 - \alpha e, \dots, v_N - \alpha e$$

where

$$e = CMAC_a(x) - (r + \gamma \max_{b \in A(y)} CMAC_b(y))$$

and $\alpha(0 < \alpha < 1)$ is a fixed learning factor.

4 Results

The environment for our CMAC-based Q-learning agents is a square in the Euclidian plane, with the endpoints $(0,0)$ and $(100,100)$, as shown in Fig. 4.

Our experiment consists of a sequence of trials. A trial begins with agents starting from the position $(56,0)$ and pushing the block, and ends when the block is taken out of the square. An agent receives some meaningful reward or penalty, when and only when the block leaves the square. Whenever an agent pass over the goal segment, a line segment between the positions $(78,100)$ and $(82,100)$, it receives a reward of 50. Whenever an agent pass over any other boundary segment of the square, it receives a penalty of -30. For any other action, the agent receives a reward of 0. Once having reached the outside of the square, the agents are relocated at the starting position, and a new trial begins again.

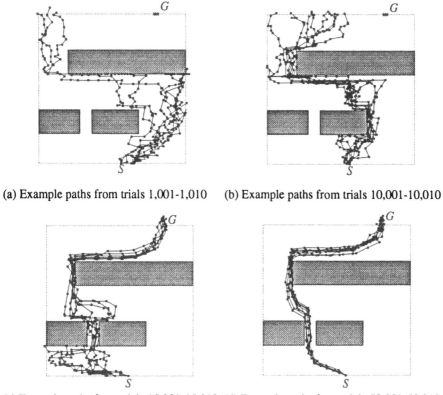

(a) Example paths from trials 1,001-1,010   (b) Example paths from trials 10,001-10,010

(c) Example paths from trials 15,001-15,010 (d) Example paths from trials 50,001-50,010

Fig. 4. Environment for learning agents and their typical paths at four different learning phases.

Given the current position x, each learning agent usually selects an action a which maximizes $Q(x, a)$, implemented by the $CMAC(a)$, but to make a random exploration it randomly selects a possible action with a probability of 0.1. While its state space remains continuous, an agent can take only discrete actions. An action is a combination of discrete force and angle taken. Possible forces and angles are m units ($m = 0, 1, \cdots, 9$) and $\frac{n\pi}{18}$ ($n = 0, 1, 2, \cdots, 18$), respectively. Each CMAC associated with a discrete action a consists of 10 tilings, each an array of 10×10 tiles, and is uniformly initialized so as to return a specific value of 100 as the initial value of $Q(x, a)$ for any state x. A learning factor α of 0.1 and a high discount factor γ of 0.9 are used.

If no noise is added to each agent's actions, the set of its possible positions is finite and does not cover the Euclidean space, because its starting position is totally fixed. So some noises are added to each agent's actions. Noises are also added to sensory inputs to each agent, to evaluate its generalization capabilities. A random noise of ± 2 is added to each agent's sensory inputs i.e. the coordinates of its positions, and others of ± 0.2 and $\pm \frac{\pi}{90}$ are added to the strength and angles of the force applied by the agent, respectively.

Typical behavior by our learning agents at their four different learning phases is shown in Fig. 4. In this case, three obstacles, each indicated by a shaded rectangle, are placed in the environment. The agents can not push the block into the area occupied by the obstacles. At initial trials, the agents leave the environment and receive penalties before successfully passing over the goal segment. They often bump the obstacles or apply a very small force at an angle of 0 or π. At trial 10,000, the agents learn to bypass the obstacles, but it is still difficult for them to reach the goal area. At trial 15,000, the agents learn to steadily reach the goal area, but the paths they take are substantially redundant at this point. This redundancy is gradually eliminated as the learning proceeds, and at trial 50,000, the agents come to effectively pass over the goal segment with a high probability.

The learning performance by our agents is evaluated quantitatively. The solid curves in figures 5, 6, and 7 respectively indicate the average percentage of successful trials, the average number of time steps per successful trial, and the average length of successful paths, each reported at every 1,000 trials. Each curve is an average over 40 independent simulation runs. The performance by those agents operating in a noiseless environment is also shown by dashed lines. Initially, as shown in these figures, our agents are not able to reach the goal segment effectively, but later they start pushing the block in a coordinated manner. Note that those agents operating in the above-mentioned noisy environment slightly outperforms those in a noiseless one.

At those points which are visited frequently, both of our agents eventually employ almost identical decision policies. Typical coordinated policies developed by our learning agents are displayed in Fig. 8, by drawing their resultant force vector at each grid of points over the state space. A vector is drawn at each regularly sampled positions when and only when each agent has a unique action with the highest utility. Each vector is scaled to indicate its relative magnitude and is drawn at the appropriate angle of the corresponding resultant force.

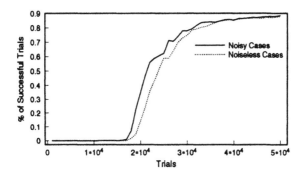

Fig. 5. Average percentage of successful trials.

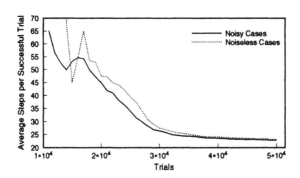

Fig. 6. Average number of time steps per trial.

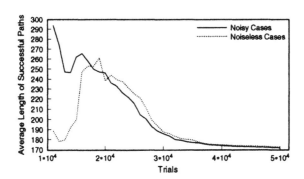

Fig. 7. Average length of successful paths.

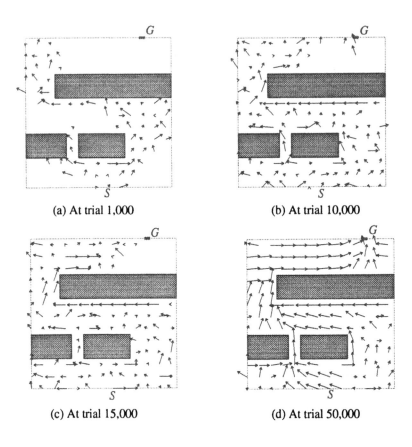

(a) At trial 1,000   (b) At trial 10,000

(c) At trial 15,000   (d) At trial 50,000

Fig. 8. Development of coordinated policies by learning agents.

As shown in the results, the learning system takes considerable time to converge, though we have not performed any fine tuning of the learning parameters. This aspect may limit the applicability of the learning scheme employed here. We plan to develop some mechanism to substantially speed up the reinforcement learning.

5   Concluding Remarks

Interesting efforts have been made to let multiple agents learn to coordinate, using various reinforcement-learning algorithms. In most of these cases, the state space for learning agents is supposed discrete. It has not been clear how effectively multiple reinforcement-learning agents are able to acquire appropriate coordinated behavior in continuous state space.

To investigate and understand the potentials of multi-agent reinforcement-learning in continuous environments, we have applied a CMAC-based Q-learning algorithm to the block pushing problem in a continuous space. Although it is not intended to model any particular real world problem and we do not say any multi-agent reinforcement-learning problems can be easily solved like this specific one, the results are encouraging.

References

1. Albus, J.S.: Brain, Behavior, and Robotics, Byte Book, Chapter 6, pp.139-179, 1981.
2. Drogoul, A., J.Ferber, B.Corbara, and D.Fresneau: A Behavioral Simulation Model for the Study of Emergent Social Structures, F.J.Varela, et al. (Eds.): Toward a Practice of Autonomous Systems: Proc. of the First European Conference on Artificial Life, The MIT Press, 1991.
3. Lin, L.-J.: Self-Improving Reactive Agents Based On Reinforcement Learning, Planning and Teaching, Machine Learning, Vol.8, 1992.
4. Ono, N., and A.T.Rahmani: Self-Organization of Communication in Distributed Learning Classifier Systems, R.F.Albrecht et al. (Eds.): Artificial Neural Nets and Genetic Algorithms: Proc. of International Conference on Artificial Neural Nets and Genetic Algorithms, Springer-Verlag Wien New York, 1993.
5. Ono, N., T.Ohira, and A.T.Rahmani: Emergent Organization of Interspecies Communication in Q-learning Artificial Organisms, in F.Móran et al.: (Eds.) Advances in Artificial Life: Proc. of the 3rd European Conference on Artificial Life, Springer, 1995.
6. Ono, N., and K.Fukumoto: Collective Behavior by Modular Reinforcement-Learning Animats, P.Maes et al.(Eds.): From Animals to Animats 4: Proc. of the 4th International Conference on Simulation of Adaptive Behavior, The MIT Press, 1996.
7. Ono, N., and K.Fukumoto: Multi-agent Reinforcement Learning: A Modular Approach, Proc. of the 2nd International Conference on Multi-agent Systems, AAAI Press, 1996.
8. Sen, S., M.Sekaran, and J.Hale: Learning to Coordinate without Sharing Information, Proc. of AAAI-94, 1994.
9. Sen, S., and M.Sekaran: Multiagent Coordination with Learning Classifier Systems, G.Weiß and S.Sen (Eds.): Adaption and Learning in Multi-agent Systems, Springer, 1996.
10. Tan, M.: Multi-agent Reinforcement Learning: Independent vs. Cooperative Agents, Proc. of the 10th International Conference on Machine Learning, 1993.
11. Yanco, H., and L.A.Stein: An Adaptive Communication Protocol for Cooperating Mobile Robots, J.-A. Meyer, et al. (Eds.): From Animals to Animats 2: Proc. of the 2nd International Conference on Simulation of Adaptive Behavior, The MIT Press, 1992.
12. Watkins, C.J.C.H.: Learning With Delayed Rewards, Ph.D.thesis, Cambridge University, 1989.

Multi-Agent Learning with the Success-Story Algorithm

Jürgen Schmidhuber and Jieyu Zhao

IDSIA, Corso Elvezia 36, CH-6900-Lugano, Switzerland
juergen@idsia.ch - http://www.idsia.ch/~juergen
jieyu@idsia.ch - http://www.idsia.ch/~jieyu

Abstract. We study systems of multiple reinforcement learners. Each leads a single life lasting from birth to unknown death. In between it tries to accelerate reward intake. Its actions and learning algorithms consume part of its life — computational resources are limited. The expected reward for a certain behavior may change over time, partly because of other learners' actions and learning processes. For such reasons, previous approaches to multi-agent reinforcement learning are either limited or heuristic by nature. Using a simple backtracking method called the "success-story algorithm", however, at certain times called evaluation points each of our learners is able to establish success histories of behavior modifications: it simply undoes all those of the previous modifications that were not empirically observed to trigger lifelong reward accelerations (computation time for learning and testing is taken into account). Then it continues to act and learn until the next evaluation point. Success histories can be enforced despite interference from other learners. The principle allows for plugging in a wide variety of learning algorithms. An experiment illustrates its feasibility.

1 Introduction / Overview

Realistic multi-agent reinforcement learning. Convergence theorems for existing reinforcement learning algorithms (e.g., [8, 18, 1, 19, 24]) require infinite sampling size as well as strong (often Markovian) assumptions about the environment. They are of great theoretical interest, but not very relevant for realistic multi-agent environments (RMAEs). RMAEs are non-Markovian: in general, no learner will have a complete model of the other learners. Each agent's environment tends to change, partly because of other learners' actions and learning processes. Also, in realistic environments, learning consumes part of each learner's limited life. So do policy tests. Furthermore, a disappointing test outcome may imply that it is already too late for some learners to collect much additional reinforcement. For instance, some learners won't be able to buy cheap shares of a company *after* the other learners drove the price up. As MERRICK FURST put it: "The biggest difference between time and space is that you can't reuse time". Obviously, realistic environments require us to rethink a bit the conventional way we measure performance.

The agents. There are n_A agents $A^1, A^2, \ldots, A^{n_A}$. They live in an unknown environment E. Each agent's life lasts from time 0 to unknown time T. Each A^i ($i \in \{1, \ldots, n_A\}$) has an *internal state* S^i and a *policy* Pol^i. Both S^i and Pol^i are variable data structures influencing probabilities of actions to be executed by A^i. Between time 0 and T, A^i repeats the following cycle over and over again ($\mathcal{A}^i$ denotes a set of possible actions): select and execute $a \in \mathcal{A}^i$ with conditional probability $P(a \mid Pol^i, S^i)$. Action a will consume time (e.g., [12, 3]) and may change E, S^i, Pol^i. (Somewhat related, but more restricted limited resource scenarios were studied in [2, 4, 5] and references therein.)

Policy modification processes. Actions in $\mathcal{A}^i$ that modify Pol^i are called "primitive learning algorithms". Certain action subsequences that include primitive learning algorithms are called "policy modification processes" (PMPs). The k-th PMP in A^i's life is denoted PMP^i_k, starts at time $s^i_k > 0$, ends at time $e^i_k < T$, $s^i_k < e^i_k$, and computes a sequence of Pol^i-modifications denoted M^i_k. Both s^i_k and e^i_k are determined by Pol^i itself during A^i's life-time by triggering special PMP-starting or PMP-ending instructions in $\mathcal{A}^i$. Since PMP execution probabilities depend on Pol^i, Pol^i influences the way it modifies itself ("self-reference").

Reward. Occasionally E provides real-valued reward. The cumulative reward obtained by A^i ($i \in \{1, \ldots, n_A\}$) in between time 0 and time $t > 0$ is denoted $R^i(t)$ (where $R^i(0) = 0$).

Agent goals. A^i's goal at some PMP-start time t is to use experience to generate Pol^i-modifications to accelerate reinforcement intake: A^i wants to let $\frac{R^i(T) - R^i(t)}{T - t}$ exceed A^i's current average speed of reinforcement intake. To determine this speed it needs a previous point $t' \leq t$ to compute $\frac{R(t) - R(t')}{t - t'}$.

Now there are two important questions. **1.** How to specify t' in a general yet reasonable way? For instance, if life consists of many successive "trials" with non-deterministic outcomes (partly depending on the other agents' behaviors), how many trials to look back into time? **2.** At time t, the expected value of $\frac{R^i(T) - R^i(t)}{T - t}$ will depend not only on Pol^i but also on $Pol^k, k \neq i$. Of all the different policies, however, A^i can explicitly modify only Pol^i. Is there at all a reasonable success criterion that A^i can achieve with its limited credit assignment powers? Both questions will be answered by the success-story criterion below.

Undoing policy modifications to achieve success histories. Each PMP-starting action in $\mathcal{A}^i$ first calls the "success-story algorithm" (SSA):
WHILE A^i's "success-story criterion" SSC^i below is not satisfied, **DO:** undo all Pol^i-modifications computed by the most recent PMP^i whose Pol^i-modifications have not been undone yet (by restoring the previous policy Pol^i — this requires storing previous values of modified Pol^i-components on a stack).

Success-story criterion (SSC). At a given PMP-start time t in A^i's life, let $Valid^i(t)$ denote the set of those previous s^i_k whose corresponding M^i_k still do exist and have not been undone yet. If $Valid^i(t)$ is not empty, then let v^i_k ($k \in \{1, 2, \ldots, V^i(t)\}$) denote the k-th valid time, ordered according to size, where $V^i(t)$ is the size of $Valid^i(t)$. SSC^i is satisfied if either $Valid^i(t)$ is empty (trivial case) or if

$$\frac{R^i(t)}{t} < \frac{R^i(t) - R^i(v_1^i)}{t - v_1^i} < \frac{R^i(t) - R^i(v_2^i)}{t - v_2^i} < \ldots < \frac{R^i(t) - R^i(v_{V^i(t)}^i)}{t - v_{V^i(t)}^i}. \quad (1)$$

SSC^i demands that each still valid time marks the beginning of a long-term reward acceleration measured up to the current time t. Each Pol^i-modification that survived SSA represents a bias shift generated by a PMP whose start marks a long-term reward speed-up. (Since probabilities of PMP-starting instructions depend on Pol^i, Pol^i can learn *when* to evaluate itself using SSA. This is important in dealing with unknown reward delays.)

SSA implementations. Using stack-based backtracking methods such as those described in [14, 16, 15, 25, 23] and section 2, one can guarantee that SSC^i will be satisfied after each new PMP^i-start, despite interference from S^i and $A^k, k \neq i$. The time consumed by backtracking itself must be taken into account by SSC^i.

SSA's generalization assumption. At the end of each SSA call, until the beginning of the next one, A^i's only temporary generalization assumption for inductive inference is: Pol^i-modifications that survived all previous SSA calls will remain useful. In absence of empirical evidence to the contrary, each still valid sequence of modifications M_i is assumed to have successfully set the stage for later valid $M_i, i > k$. In unknown environments, which other generalization assumption would make sense? Since life is one-way (time is never reset), during each SSA call A^i has to generalize from a *single* experience concerning the usefulness of policy modifications executed after any given previous point in time: the average reward per time since then.

When will SSC be satisfiable in a non-trivial way? In irregular and random environments there is no way of justifying permanent Pol^i-modifications by SSC^i. Also, a trivial way of satisfying SSC^i is to never make a Pol^i-modification. Let us assume, however, that E, S^i and A^i (representing the system's initial bias) do indeed allow for Pol^i-modifications triggering long-term reward accelerations. (This is an instruction set-dependent assumption much weaker than the typical Markovian assumptions made in previous RL work, e.g., [8, 18, 19, 24].) Now, if we prevent all instruction probabilities from vanishing (see concrete implementation in section 2), then Pol^i cannot help triggering Pol^i-modifications occasionally, and keeping those consistent with SSC^i. In this sense, A^i cannot help getting better.

Measuring effects of learning on later learning. Since some Pol^i-modification's success recursively depends on the success of later Pol^i-modifications for which it sets the stage, SSA (unlike previous RL methods) automatically provides a basis for "learning how to learn". Note that the older some modification, the more time will have passed to collect experiences with its long-term consequences, and the more stable it will be if it is indeed useful and not just there by chance.

SSA's limitations. Of course, in general environments neither SSA nor any other scheme is guaranteed to find "optimal" policies that will lead to maximal cumulative reward for each agent (we'll never know for sure which stocks will go up). SSA is only guaranteed to selectively undo those Pol^i-modifications that were not empirically observed to lead to an overall speed-up of A^i's average reward intake. Still, this is more than can be said about existing, interesting reinforcement learning algorithms (see, e.g., [8, 18, 19, 24, 6, 7, 11]) or multi-agent learning algorithms (see, e.g., [21, 20, 17]).

Outline. The next section will describe a concrete implementation of the principles above. We will see how each agent's policy can be represented by a set of variable probability distributions on a set of assembler-like instructions, how the policy can build the basis for generating and executing a lifelong instruction sequence, and how each policy can modify itself executing sequences of policy-modifying instructions. We will also explain the details of the SSA-based reinforcement acceleration method.

2 Concrete Implementation

In what follows, we will describe details of a single agent implementing the principles presented in section 1. Its cumulative reward obtained in between time 0 and time $t > 0$ is denoted $R(t)$, where $R(0) = 0$. Several such agents will interact in the experiment in section 3.

Architecture. There are m addressable *program cells* with addresses ranging from 0 to $m - 1$. The variable, integer-valued contents of the program cell with address i are denoted d_i. An internal variable *Instruction Pointer* (*IP*) with range $\{0, \ldots, m - 1\}$ always points to one of the program cells (initially to the first one). Actions (or instructions) and their parameters are encoded by a fixed set I of n_{ops} integer values $\{0, \ldots, n_{ops} - 1\}$. For each value j in I, there is an instruction B_j with N_j integer-valued parameters.

Stochastic policy. For each program cell i there is a variable probability distribution Pol_i on I. For every possible $j \in I$, $(0 \leq j \leq n_{ops}-1)$, Pol_{ij} specifies for cell i the conditional probability that, when pointed to by *IP*, its contents will be set to j. The set of all current Pol_{ij}-values defines a probability matrix Pol (the learner's *current, stochastic policy*) with columns Pol_i $(0 \leq i \leq m - 1)$. If $IP = i$, then the contents of i, namely d_i, will be interpreted as instruction B_{d_i}, and the contents of cells immediately following i will be interpreted as B_{d_i}'s arguments, to be selected according to the corresponding Pol-values.

Normal instructions. Normal instructions are those that don't change the policy. An example is the following one (later we will introduce many additional, problem-specific instructions):

JumpHome — set $IP = 0$ *(jump back to 1st program cell).*

Primitive learning algorithms. By definition, a primitive learning algorithm is an instruction that modifies Pol. In what follows, the symbols $a1, a2, a3$ stand for instruction parameters that may take on integer values between 0 and $n_{ops} - 1$. For the moment, we will consider only one primitive learning algorithm called *IncProb* (in conjunction with other instructions, *IncProb* may be used as part of more complex learning processes):

IncProb(a1, a2, a3) — Set $i := a1 * n_{ops} + a2$ and $j := a3$. If $0 \leq i \leq m - 1$, increase Pol_{ij} by γ percent, and renormalize Pol_i (but prevent P-values from falling below a minimal value ϵ, to avoid near-determinism). In the experiments, we use $\gamma = 15, \epsilon = 0.001$.

Evaluation instruction. The learner can end a PMP and trigger an evaluation of its own lifelong performance so far, by executing the instruction *"PrepareEvaluation"*.

PrepareEvaluation() — End the current PMP if there is one: temporarily disable learning instructions (e.g., *IncProb*) by preventing them from executing further policy modifications, until $NumR$ additional, non-zero reinforcement signals have been received (this will trigger a new PMP starting with an evaluation point and a SSA call — see basic cycle below). In the experiment we will set $NumR = 5$.

Initialization. At time 0 (system birth), we initialize policy Pol with maximum entropy distributions (all Pol_{ij} equal), and set $IP = 0$. We introduce an initially empty stack S that allows for variable-sized stack entries, and the conventional *push* and *pop* operations. The integer variable $NumR$ (initially zero) is used to count down the remaining non-zero reinforcement signals since the last execution of *"PrepareEvaluation"*. The Boolean variable $PolChanged$ is initially false.

Basic cycle of operations. Starting at time 0, until time T (system death), the system repeats the following basic instruction cycle over and over again (while time is continually increasing):

1. Randomly generate an integer $j \in I$ according to matrix column Pol_{IP} (the probability distribution of the program cell pointed to by IP). Set program cell content $d_{IP} := j$. Translate j into the corresponding current instruction B_j. Look up the number N_j of cells required to store B_j's parameters.
   IF $IP > m - N_j - 2$, **THEN** set IP to 0, go to step **1**.
   ELSE generate instruction arguments for the N_j cells immediately following IP, according to their probability distributions $Pol_{IP+1}, ..., Pol_{IP+N_j}$, and set $IP \leftarrow IP + N_j + 1$.
2. **IF** B_j is a primitive learning algorithm, and not currently disabled by a previous *"PrepareEvaluation"* instruction, **THEN** push copies of those Pol_i to be modified by B_j onto S.
3. Execute instruction B_j. This may change (1) environment, (2) IP, (3) Pol itself. In case (3), set the Boolean variable $PolChanged = TRUE$. **IF** B_j is *"PrepareEvaluation"* and $PolChanged = TRUE$, **THEN** set $NumR$ equal to B_j's actual parameter plus one.
4. **IF** $NumR > 0$ and non-zero reinforcement occurred during the current cycle, **THEN** decrement $NumR$. **IF** $NumR = 0$, **THEN** call SSA to enforce the learner's SSC by backtracking:

SSA.1. Set variable t equal to current time.
   IF there is no *"tag"* stored somewhere in S (tags are pairs of *valid* times (see section 1) and corresponding cumulative reinforcements pushed in earlier executions of step 4), **THEN** push the tag $(t, R(t))$ onto S, set

$PolChanged = FALSE$, and go to **1** (this ends the current evaluation interval).

SSA.2. Denote the topmost tag in S by $(t', R(t'))$. **IF** there are no further tags, **THEN** set variable $t'' = 0$ (recall $R(t'') = R(0) = 0$). **ELSE** denote the last but topmost tag in S by $(t'', R(t''))$.

SSA.3. IF $\frac{R(t)-R(t')}{t-t'} > \frac{R(t)-R(t'')}{t-t''}$ **THEN** push $(t, R(t))$ as a new tag, set $PolChanged = FALSE$, and go to **1**. This ends the current evaluation interval.

ELSE pop off all stack entries above the one for tag $(t', R(t'))$ (these entries will be former policy components saved during an earlier execution of step 2), and use them to restore Pol as it used to be before time t'. Then also pop off the tag for $(t', R(t'))$. Go to **SSA.1**.

Comment: each step above will consume various amounts of system life-time.

3 Experiment: Predator and Prey

We have conducted numerous successful experiments with systems like the one described above. See, e.g., [15, 14, 16, 23, 25]. For instance, in [16] two agents A and B learn to cooperate in a 600×500 pixel environment with obstacles. They learn to solve a complex task that requires (1) agent A to find and take a key "key A"; (2) agent A go to a door "door A" and open it for agent B; (3) agent B to enter through "door A", find and take another key "key B"; (4) agent B to go to another door "door B" to open it (to free the way to the goal); (5) one of the agents to reach the goal. Reward is provided only if one of the agents touches the goal — there is no intermediate reinforcement. In the beginning, the goal is found only every 300,000 basic cycles. Through self-modifications and SSA, however, within 130,000 trials the average trial length decreases by a factor of 60. Our method indeed forces both agents to cooperate to accelerate reward intake.

We also implemented another multi-agent system: a fully recurrent net whose agents are its connections. Each connection's current weight represents its current policy. The net receives inputs from the environment through its input units, and controls an "animat" via output patterns across its output units. The animat lives in a simple maze. Whenever a goal is found, there is constant reinforcement 1.0, and the animat is reset to its start position. For each connection, there is a stack storing the information required for its SSA calls. After about 10,000 "trials", average trial length improved by a factor of about 30. The system also found a shortest path. In the end, no weight stack had more than 7 entries (each followed by faster reinforcement intake than all the previous ones). Details can be found in [15].

The illustrative experiment in this section, however, provides a different example where there is no cooperation but only competition. A system consisting of three co-evolving agents chasing each other learns rather sophisticated, stochastic predator and prey strategies. Despite the fact that each agent's task gets more and more difficult over time due to continually improving opponents, each agent is able to establish non-trivial, lifelong histories of reward accelerations. See also [25].

Task (see also [25]). There are three co-evolving, IS-based agents A, B, C. Each agent simultaneously is both predator and prey. A's prey is B. B's prey is C. C's prey is A. A's predator is C. B's predator is A. C's predator is B. Each agent's goal is to catch its prey as often (and as quickly) as possible, while evading its predator. Each time it catches its prey, it receives a reinforcement of 1.0; each time it gets caught, it receives negative reinforcement -1.0 and is randomly reset to a new position. Each agent's environment changes because the other agents in its environment learn and change. We have a symmetric zero-sum game, resulting in a continuous evolutionary race. Still, each agent is able to establish non-trivial, lifelong histories of reward accelerations.

The simulated environment consists of an area of 600×800 pixels. See Figure 1. Obstacles are set in advance but can be changed by the experimenter at any later time. Each agent has circular shape and a diameter of 30 pixel widths. At a given time, it can move up to 9 steps in 8 different directions. Directions are represented as integers in $\{0,\ldots,7\}$: 0 for north, 1 for northeast, 2 for east,... etc.. A single step places the agent's center at the closest pixel 5 pixel widths away in the current direction. Each agent is equipped with limited "active" sight: by executing certain actions, it can sense obstacles or its prey or its predator within up to 10 steps in front of it. It also can make turns relative to prey and predator position. Since each agent can reach any position in the field (except those blocked by obstacles), the multi-agent system's entire state space is *huge*. Each agent's instruction set includes the following $n_{ops} = 10$ instructions:

B_0: *Move(n)* — if $0 \le n \le 9$ then move n steps forward in the current direction.

B_1: *Turn(d)* — If $0 \le d \le 7$ then change current direction D to $(D+d+4)mod8$.

B_2: *TurnRelativeToPredator(d)* — If $0 \le d \le 7$ then turn to the direction that best matches the line connecting the centers of agent and its predator, then *Turn(d)*.

B_3: *TurnRelativeToPrey(d)* — analogous to B_2.

B_4: *LookForPredator(n)* — if the agent's predator is not within $n + 1$ steps in front of the agent $(0 \le n \le 9)$, then increase IP by 4 (this is a limited kind of conditional jump).

B_5: *LookForPrey(n)* — analogous to B_4.

B_6: *LookForObstacle(n)* — analogous to B_4.

B_7: *JumpHome* — jump back to program cell 0 (see section 2.)

B_8: *PrepareEvaluation()* — see "evaluation instruction" in section 2.

B_9: *IncProb(a1, a2, a3)* : See "primitive learning algorithm" in section 2.

Each instruction occupies 2 successive program cells (the second one unused if there are no parameters) except for *IncProb*, which occupies 4. We set $m = 50$.

Given the primitives above, each agent faces a complex partially observable Markov decision problem (POMDP), e.g., [13, 22, 6, 7, 11, 10, 9, 23] — the current input by itself does not necessarily provide all information needed to determine the optimal next action. The agents may use their instruction pointers to disambiguate inputs.

SSA calls. For each agent, an evaluation point (and SSA call) occurs after each 5th consecutive non-zero reinforcement signal following each execution of instruction *PrepareEvaluation* (i.e., $NumR = 5$).

Copying successful policies. To achieve a balanced evolutionary race with more or less equal opponents, we let agents with inferior performance occasionally "steal" the best agent's policy. The corresponding copying procedure is initiated at special times called "copypoints". Once a copypoint occurs, we wait until all agents have encountered their respective next evaluation points and have finished the corresponding SSA calls; until then, no agent may start another PMP by itself (*IncProb* disabled). Then we make each agent's policy equal to the policy of the currently most successful agent among the three. The proper way of doing this is to act as if the corresponding sequence of modifications caused to the policies of the two less successful agents was generated by standard PMPs — the old probability distributions are pushed onto the respective stacks, together with information about the current time and the reinforcement so far. Copypoints occur every 10^6 instruction cycles.

Results. After each agent has executed 10^9 instruction cycles, all three agents (whose behavior can be observed directly on the screen) exhibit quite interesting pursuit-evasion behaviors. They all chase each other, using existing obstacles trying to avoid being caught. The final number of valid probability modifications per agent is reflected by the final stack size, which is about 350. Figure 2 shows a snapshot with recent movement traces, but unfortunately the often surprising and complex dynamics cannot be conveyed by a single snapshot. Since the game is open-ended, no agent reduces the frequency of its self-modifications but keeps learning and reacting to changes of the others' behavior. The agents' strategies appear hard to analyze. In particular, although each agent quickly replaces the initial high-entropy probability distributions of its policy by low-entropy distributions, it often leaves significant probability values for alternative behavioral sequences. Much of each agent's policy is not committed to any obvious deterministic strategy, but is stochastic instead.

However, there are policy fragments making sense to a human observer. For instance, Figure 2 depicts a policy fragment that makes the following action sequence likely to occur:

1. *If there is no obstacle within 6 steps ahead, then skip instructions 2 and 3.* **2.** *Execute a random instruction (whose parameter is likely to be 5).* **3.** *Face the prey (the parameter determining relative direction to the prey is set to zero).* **4.** *Move 8 steps forward.* **5.** *If the prey cannot be seen within 9 steps, then skip instructions 6 and 7.* **6.** *If (at the next time step) the prey cannot be seen within 6 steps ahead, then skip instructions 7 and 8 (in conjunction with instruction 5, instruction 6 may be used to test whether the prey is getting closer from the vision periphery).* **7.** *Move 8 steps forward (this instruction will be executed, e.g., if the prey seems close or getting closer).* **8.** *If the predator cannot be seen within 9 steps ahead, then skip instructions 9 and 10.* **9.** *If the predator cannot be seen within 8 steps ahead, then skip instructions 10 and*

11 (in conjunction with instruction 8, instruction 9 may be used to test whether the predator is getting closer from the vision periphery). 10. Face the prey (this instruction will typically turn the agent away from the predator — but note that the behavior of the agent's prey and its predator also depend on each other in a complicated manner). 11. Jump back to program cell 0. 12. ... a lot of additional, partly highly deterministic code.

Although this fragment does not exactly look like an exercise in elegant programming, it represents part of a strategy that works reasonably well for a broad range of situations.

How to achieve success stories in zero sum games? The sum of all rewards occurring during the predator and prey game is zero. Each agent's task gets more and more difficult over time. So how can each agent create a non-trivial history of policy modifications, each corresponding to a lifelong reward acceleration? The answer is: each agent collects a lot of negative reinforcement during its life, and actually comes up with a history of policy modifications causing less and less negative cumulative long-term rewards. That's why the stacks of all agents tend to grow continually.

4 Conclusion

Each agent's credit assignment powers are limited in the sense that it can change only its own policy, not the ones of the other agents. Still, each call of each agent's success-story algorithm will establish a history of valid self-modifications representing a success story. Each surviving self-modification will correspond to a long-term reward acceleration.

Acknowledgments
Jieyu Zhao is supported by SNF grant 21-43'417.95 "Incremental Self-Improvement".

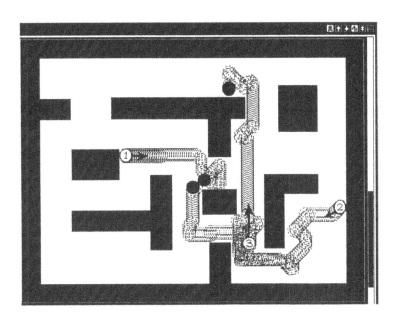

Fig. 1. *Snapshot taken during the pursuit-evasion game. Agents A, B, C come from initial positions 1, 2, 3, respectively. Arrows indicate initial directions of each agent. Agent C successfully escapes its predator B by quickly moving north, partly because B turns to avoid contact with its predator A, which was on its way east towards B but then turned away when its predator C passed by. This turn provides additional benefits for A, because of B's own turn: while heading in the general direction of C, B gets caught by A.*

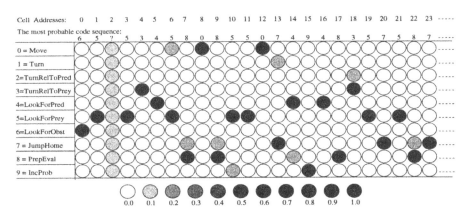

Fig. 2. *Columns 0-23 of the final probability matrix for the predator/prey game. It was computed by self-modification sequences generated according to the probability matrix itself (initially, all probability distributions were maximum entropy distributions). We may say that, at a given time in system life, the probability matrix embeds its own current stochastic learning strategy, and also its own "meta-learning" strategy, etc..*

References

1. A. G. Barto. Connectionist approaches for control. Technical Report COINS 89-89, University of Massachusetts, Amherst MA 01003, 1989.
2. D. A. Berry and B. Fristedt. *Bandit Problems: Sequential Allocation of Experiments*. Chapman and Hall, London, 1985.
3. M. Boddy and T. L. Dean. Deliberation scheduling for problem solving in time-constrained environments. *Artificial Intelligence*, 67:245–285, 1994.
4. J. C. Gittins. *Multi-armed Bandit Allocation Indices*. Wiley-Interscience series in systems and optimization. Wiley, Chichester, NY, 1989.
5. R. Greiner. PALO: A probabilistic hill-climbing algorithm. *Artificial Intelligence*, 83(2), 1996.
6. T. Jaakkola, S. P. Singh, and M. I. Jordan. Reinforcement learning algorithm for partially observable Markov decision problems. In G. Tesauro, D. S. Touretzky, and T. K. Leen, editors, *Advances in Neural Information Processing Systems 7*, pages 345–352. MIT Press, Cambridge MA, 1995.
7. L.P. Kaelbling, M.L. Littman, and A.R. Cassandra. Planning and acting in partially observable stochastic domains. Technical report, Brown University, Providence RI, 1995.
8. P. R. Kumar and P. Varaiya. *Stochastic Systems: Estimation, Identification, and Adaptive Control*. Prentice Hall, 1986.
9. M.L. Littman, A.R. Cassandra, and L.P. Kaelbling. Learning policies for partially observable environments: Scaling up. In A. Prieditis and S. Russell, editors, *Machine Learning: Proceedings of the Twelfth International Conference*, pages 362–370. Morgan Kaufmann Publishers, San Francisco, CA, 1995.
10. R. A. McCallum. Overcoming incomplete perception with utile distinction memory. In *Machine Learning: Proceedings of the Tenth International Conference*. Morgan Kaufmann, Amherst, MA, 1993.
11. M. B. Ring. *Continual Learning in Reinforcement Environments*. PhD thesis, University of Texas at Austin, Austin, Texas 78712, August 1994.
12. S. Russell and E. Wefald. Principles of Metareasoning. *Artificial Intelligence*, 49:361–395, 1991.
13. J. Schmidhuber. Reinforcement learning in Markovian and non-Markovian environments. In D. S. Lippman, J. E. Moody, and D. S. Touretzky, editors, *Advances in Neural Information Processing Systems 3*, pages 500–506. San Mateo, CA: Morgan Kaufmann, 1991.
14. J. Schmidhuber. A general method for multi-agent learning in unrestricted environments. In *Adaptation, Co-evolution and Learning in Multiagent Systems, Technical Report SS-96-01*, pages 84–87. American Association for Artificial Intelligence, Menlo Park, Calif., 1996.
15. J. Schmidhuber. A general method for incremental self-improvement and multi-agent learning in unrestricted environments. In X. Yao, editor, *Evolutionary Computation: Theory and Applications*. Scientific Publ. Co., Singapore, 1997. In press.
16. J. Schmidhuber, J. Zhao, and M. Wiering. Simple principles of metalearning. Technical Report IDSIA-69-96, IDSIA, 1996.
17. S. Sen, editor. *Adaptation, Co-evolution and Learning in Multiagent Systems, Papers from the 1996 AAAI Symposium, Technical Report SS-96-01*. American Association for Artificial Intelligence, Menlo Park, Calif., 1996.
18. R. S. Sutton. Learning to predict by the methods of temporal differences. *Machine Learning*, 3:9–44, 1988.

19. C. J. C. H. Watkins and P. Dayan. Q-learning. *Machine Learning*, 8:279–292, 1992.

20. G. Weiss, editor. *Learning in Distributed Artificial Intelligence Systems, ECAI-96 Workshop Notes, Budapest, Hungary.* 1996.

21. G. Weiss and S. Sen, editors. *Adaption and Learning in Multi-Agent Systems.* LNAI 1042, Springer, 1996.

22. S.D. Whitehead. *Reinforcement Learning for the adaptive control of perception and action.* PhD thesis, University of Rochester, February 1992.

23. M.A. Wiering and J. Schmidhuber. Solving POMDPs with Levin search and EIRA. In L. Saitta, editor, *Machine Learning: Proceedings of the Thirteenth International Conference*, pages 534–542. Morgan Kaufmann Publishers, San Francisco, CA, 1996.

24. R. J. Williams. Simple statistical gradient-following algorithms for connectionist reinforcement learning. *Machine Learning*, 8:229–256, 1992.

25. J. Zhao and J. Schmidhuber. Incremental self-improvement for life-time multi-agent reinforcement learning. In Pattie Maes, Maja Mataric, Jean-Arcady Meyer, Jordan Pollack, and Stewart W. Wilson, editors, *From Animals to Animats 4: Proceedings of the Fourth International Conference on Simulation of Adaptive Behavior, Cambridge, MA*, pages 516–525. MIT Press, Bradford Books, 1996.

On the Collaborative Object Search Team: A Formulation

Yiming Ye and John K. Tsotsos*

Department of Computer Science
University of Toronto
Toronto, Ontario, Canada M5S 1A4

Abstract. This paper gives a formulation of a collaborative object search team and studies the learning, interaction and organization within this multiagent environment. Each team member is assumed to be a mobile platform equipped with an active camera and recognition algorithms so that images of the environment can be taken and analyzed for the target object. The goal of the team is to find the target within a given time constraint. In order to do this, the agents must interact and collaborate with each other and must learn and modify the various cooperation styles based on the search results.

1 Introduction

Many researchers in the field of Distributed Artificial Intelligence are beginning to build agents that can work in a complex, dynamic multiagent domains[13]. Such domains include virtual theater[1], realistic virtual training environments [10] [11] [13], RoboCup robotic and virtual soccer [6] and robotic collaboration by observation [7].

This paper focuses on our recent research effort aimed at developing theories and systems for multiagent object search —— the task of searching for a $3D$ object in a $3D$ environment by a group of robots. Constructing multiagent object search systems requires facing many hard research challenges, such as goal-driven behavior, reactivity, real-time performance, planning, learning, coordination, and spatial and temporal reasoning. We believe that given its *real-world nature*, the multiagent object search task reflects many of the key characteristics of other real-world tasks in a dynamic multiagent environment.

* The authors are grateful to Eric Harley and the two reviewers of the ICMAS-96 workshop on Learning, Interactions and Organizations in Multiagent Environment for their valuable comments. This work is supported by a research fellowship from IBM Centre For Advanced Studies and by the ARK (Autonomous Robot for a Known environment) Project, which receives its funding from PRECARN Associates Inc., Industry Canada, the National Research Council of Canada, Technology Ontario, Ontario Hydro Technologies, and Atomic Energy of Canada Limited. John K. Tsotsos is a fellow of the Canadian Institute for Advanced Research. Yiming Ye would like to thank Karen Bennet for her help.

Here we study the problem of object search by a team of *collaborative* robotic agents. By collaborative we mean that the agents are working together collaboratively in order to realize a common goal —— finding the target. A multi-agent object search system is quite different from the task of object search by a single robotic agent, on which we have done an extensive research and experiments [15] [16]. The multiagent team activities are not merely a union of simultaneous, independent individual activities —— each agent is not merely searching alone without considering the actions of other agents. Focusing on the multiagent aspect of the object search task brings to attention a number of challenging issues where learning is involved, such as:

- Interaction: How do agents communicate? How do agents coordinate in a team in order to find the target in the environment as early as possible? How should the patterns of interaction that characterize coordinated behavior be modeled, and how do agents learn from the effectiveness of the previous patterns of interaction?
- Knowledge representation and organization: How do search agents represent their local views of the world? How is the local knowledge updated or learned as a consequence of the agent's own action? How do search agents represent their local views of other agents? How do agents revise and learn beliefs about other agents based on exchanged information? How do agents organize knowledge about self, other agents, and the world such that newly learned facts can be easily integrated into the representation?
- Planning and Reasoning: How do agents plan their actions based on the knowledge about themselves, the knowledge about other agents, the knowledge learned from the interaction during the search process, and the knowledge learned from the effectiveness of its geometric relationships with other agents?

In this paper, we first formulate the task of multiagent object search and analyze its structure and complexity, then we study various issues concerning learning.

2   The Object Search Agent

In this section, we describe some concepts about the object search agent in the multiagent environment, such as the $3D$ region to be searched, the model of the search agent, the state parameters of a search agent, the search agent's local knowledge about the world, the operation of a search agent and its cost. We assume throughout this paper that there are a total of m search agents a_1, a_2, ..., a_m available in the object search team.

The search region Ω can be of any form, and it is assumed that we know the boundary of Ω and its internal geometric configuration exactly. In practice, we tessellate the region Ω into a series of elements c_i: $\Omega = \bigcup_{i=1}^{n} c_i$ and $c_i \bigcap c_j = 0$ for $i \neq j$. We call each of the element c_i a cell of the environment and we assume in this paper that there are a total of n cells. In addition to this, we introduce another "cell" c_{out} to refer to the region that is outside the search region Ω. Usually, the search region is an office-like environment, and it is tessellated into little cubes of equal size. The goal of the search team is to determine which cube c_i $(1 \leq i \leq n, out)$ contains the target center. When we say c_{out} contains the target center, we mean that the target is outside the region Ω.

The model of the search agent (Fig. 1(a)(b)) is taken to be a mobile platform with a robotic head and a camera that can pan, tilt and zoom (Note: this model is based on the ARK robot and the Laser Eye [9]). The camera's image plane is assumed to be always coincident with its focal plane.

The state s_a of a search agent a is uniquely determined by 7 parameters $(x_a, y_a, z_a, w_a, h_a, p_a, t_a)$, where (x_a, y_a, z_a) is the position of the camera center; w_a, h_a are the width and height of the solid viewing angle of the camera; and p_a, t_a are the the camera's viewing direction (Fig. 2). The position (x_a, y_a, z_a) can be adjusted by moving the mobile platform. In the case of a mobile platform, z_a is a fixed constant (the height of the camera), thus, only x_a and y_a are adjustable. The viewing angle size $\langle w_a, h_a \rangle$ can be adjusted by the zoom lens of the camera. Pan and tilt $\langle p_a, t_a \rangle$ can be adjusted by the motors on the robotic head.

The agent's knowledge about the possible target position is specified by a probability distribution function $\mathbf{p}$. The term $\mathbf{p}(a_i, c_j, \tau)$ gives the belief of agent a_i regarding the probability that the center of the target is within cell c_i at time τ. Before the team search process begins, each agent should obtain its own probability distribution function for the search region. This represents the agent's local knowledge about the world. This knowledge should satisfy the following constraint,

$$\sum_{j=1}^{n} \mathbf{p}(a_i, c_j, \tau_{0-}) + \mathbf{p}(a_i, c_{out}, \tau_{0-}) = 1 \ , \tag{1}$$

where τ_{0-} means the time before the search process, and $\mathbf{p}(a_i, c_{out}, \tau_{0-})$ refers to agent a_i's belief that the target is outside the search region before the search process. If agent a_i is not able to determine whether the target is more likely to be outside the search region or inside the search region, then $\mathbf{p}(a_i, c_{out}, \tau_{0-}) = 0.5$. If agent a_i has no knowledge about the possible target distribution within the search region, then it assumes a uniform distribution

$$\mathbf{p}(a_i, c_j, \tau_{0-}) = \frac{1 - \mathbf{p}(a_i, c_{out}, \tau_{0-})}{n} \ , \tag{2}$$

for $j = 1, \ldots n$.

An operation $\mathbf{f}(a_i, s_{a_i}, r_{a_i}^{(j)})$ for agent a_i entails two steps: (1) take a *perspective* projection image according to state s_{a_i}, and then (2) search the image for the target using the recognition algorithm $r_{a_i}^{(j)}$. We assume that each agent can have several recognition algorithms that can be used to detect the target; $r_{a_i}^{(j)}$ means agent a_i's jth recognition algorithm.

The cost $\mathbf{t}(\mathbf{f})$ for an action $\mathbf{f} = \mathbf{f}(a_i, s_{a_i}, r_{a_i}^{(j)})$ gives the total time needed for a_i to execute the action $\mathbf{f}(a_i, s_{a_i}, r_{a_i}^{(j)})$. It includes (1) time to manipulate the hardware to the status s_{a_i} specified by $\mathbf{f}$; (2) time to take a picture using the camera on a_i; (3) time to update a_i's inner representation of the world; and (4) time to run the recognition algorithm $r_{a_i}^{(j)}$ specified by $\mathbf{f}$. We assume that: (2) and (3) are same for all the actions; (4) is known for any recognition algorithm and is constant for a given recognition algorithm.

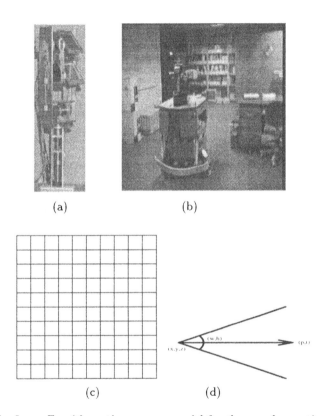

(a)   (b)

(c)   (d)

Fig. 1. (a) The Laser Eye (the active camera model for the search agent). At the top is the Optech laser range finder; at the bottom is the zoom and focus controlled lens. The two mirrors are used to ensure collinearity of effective optical axes of the camera lens and the range finder. The pan-tilt unit is operated by two DC servocontrolled motors from Micromo Electronics, equipped with gearboxes and optical encoders. (b) The ARK mobile platform (the mobile platform for the search agent). It is a 3 wheeled Cybermotion K2A platform equipped with a small number of sonar transducers. The Laser Eye can be mounted on a robotic head on the platform. The ARK mobile platform can move freely in an office-like environment or a known industrial environment. (c) 2D illustration of the tessellation of the search region Ω. (d) 2D illustration of the state s_a of a search agent a. Where (x, y, z) are the coordinates of the camera center, (w, h) is the viewing angle size of the camera, and (p, t) is the viewing direction of the camera.

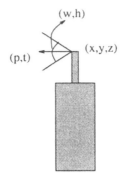

Fig. 2. A 2D illustration of the state of the hardware. Coordinate (x, y, z) indicate the position of the center of the camera; (w, h) is the viewing angle size of the camera; and (p, t) is the viewing direction of the camera.

3 Awareness and Knowledge Adaptation

Awareness means that the agent has knowledge about itself and the world. Knowledge adaptation means that the agent perceives the environment through actions and incorporates the results of perception into its own internal representation.

Each agent a_i has inner knowledge about its ability to detect the target by applying a given action $\mathbf{f}$. The detection ability is represented by the detection function $\mathbf{b}(a_i, c_j, \mathbf{f})$, which gives the conditional probability that agent a_i will detect the target given that the center of the target is located within cell c_j, and the operation is $\mathbf{f}$. This function is approximated by assuming that the target center is located *at the center* of c_j. For any operation, if the projection of the center of the cell c_j is outside the image, we assume $\mathbf{b}(a_i, c_j, \mathbf{f}) = 0$; if the cell is occluded or too far from the camera or too near to the camera, we also have $\mathbf{b}(a_i, c_j, \mathbf{f}) = 0$. In general [17], $\mathbf{b}(a_i, c_j, \mathbf{f})$ is determined by various factors, such as illumination intensity, occlusion, and orientation, etc. The value of the detection function $\mathbf{b}(a_i, c_j, \mathbf{f})$ for operation $\mathbf{f}$ can be obtained by transforming from a pre-recorded standard detection function for the recognition algorithm used by $\mathbf{f}$.

It is obvious that agent a_i's knowledge about the probability of detecting the target by applying action $\mathbf{f}$ by itself is given by

$$P(\mathbf{f}) = \sum_{j=1}^{n} \mathbf{p}(a_i, c_j, \tau_{\mathbf{f}}) \mathbf{b}(a_i, c_j, \mathbf{f}) \ , \qquad (3)$$

where $\tau_{\mathbf{f}}$ is the time just before $\mathbf{f}$ is applied. It is obvious that if a cell c_j is outside the field of view of the camera determined by $\mathbf{f}$ or is occluded with respect to the geometry specified by $\mathbf{f}$, then $\mathbf{b}(a_i, c_j, \mathbf{f}) = 0$. We use

$$\Omega(\mathbf{f}) = \left\{ c \mid c \in \Omega \wedge \mathbf{b}(a_i, c, \mathbf{f}(a_i, s_{a_i}, r_{a_i}^{(j)})) \neq 0 \right\} \qquad (4)$$

to represent those cells whose detection function values are not zero with respect to a_i and $\mathbf{f}$. $\Omega(\mathbf{f})$ is called the *influence range* for $\mathbf{f}$. Of course,

$$P(\mathbf{f}) = \sum_{c \in \Omega(\mathbf{f})} \mathbf{p}(a_i, c, \tau_{\mathbf{f}}) \mathbf{b}(a_i, c, \mathbf{f}) . \tag{5}$$

For any agent a_i, its beliefs on the possible target positions are represented by $\mathbf{p}(a_i, c, \tau)$, and these beliefs change over time as the agent perceives the world. If an action by agent a_i finds the target, then a_i will report the result and the task is accomplished. If an action by agent a_i fails to detect the target, then a_i needs to incorporate this result into its *local* knowledge representation regarding the target distribution. This is the process of *learning from action*. The agent uses Bayes' formula to update its *local* knowledge base. Let α_j be the event that the center of the target is in cell c_j, and α_{out} be the event that the center of the target is outside the search region. Let β be the event that after applying a recognition action $\mathbf{f}$ by agent a_i, the recognizer successfully detects the target. Then $P(\neg\beta \mid \alpha_j)$ gives the probability of not detecting the target by action $\mathbf{f}$ of agent a_i given that the target center is within cube c_j. Since $\mathbf{b}(a_i, c_j, \mathbf{f})$ gives the probability of detecting the target by action $\mathbf{f}$ given that the target center is within cube c_j. We have

$$P(\neg\beta \mid \alpha_j) = 1 - \mathbf{b}(a_i, c_j, \mathbf{f}) . \tag{6}$$

It is obvious that $P(\alpha_j \mid \neg\beta)$ gives the probability that the target center is within the cube c_j given that $\mathbf{f}$ failed to detect the target. If we represent the updated probability for cube c_j as $\mathbf{p}(a_i, c_j, \tau_{\mathbf{f}_+})$, where $\tau_{\mathbf{f}_+}$ is the time after $\mathbf{f}$ is applied. Then we should have

$$\mathbf{p}(a_i, c_j, \tau_{\mathbf{f}_+}) = P(\alpha_j \mid \neg\beta) . \tag{7}$$

Since the events $\alpha_1, \ldots, \alpha_n, \alpha_{\text{out}}$ are mutually complementary and exclusive, from Bayes' formula, we get

$$P(\alpha_i \mid \neg\beta) = \frac{P(\alpha_i) P(\neg\beta \mid \alpha_i)}{\sum_{j=1}^{n,\text{out}} P(\alpha_j) P(\neg\beta \mid \alpha_j)} , \tag{8}$$

for $i = 1, \ldots, n, \text{out}$

Thus, by replacing $P(\alpha_i \mid \neg\beta)$ with $\mathbf{p}(a_i, c_j, \tau_{\mathbf{f}_+})$, $P(\alpha_i)$ with $\mathbf{p}(a_i, c_j, \tau_{\mathbf{f}})$, and $P(\neg\beta \mid \alpha_i)$ with $1 - \mathbf{b}(a_i, c_j, \mathbf{f})$, we obtain the following probability updating rule

$$\mathbf{p}(a_i, c_j, \tau_{\mathbf{f}_+}) \leftarrow \frac{\mathbf{p}(a_i, c_j, \tau_{\mathbf{f}}) \left(1 - \mathbf{b}(a_i, c_j, \mathbf{f})\right)}{\sum_{k=1}^{n,\text{out}} \mathbf{p}(a_i, c_k, \tau_{\mathbf{f}}) \left(1 - \mathbf{b}(a_i, c_k, \mathbf{f})\right)} , \tag{9}$$

where $j = 1, \ldots, n, \text{out}$.

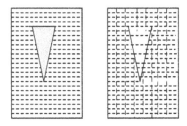

Fig. 3. A 2D illustration of the belief updating process. When an agent applies an action in the environment but fails to detect the target, it's belief on the target distribution within the influence range of the action is decreased and its belief on the target distribution outside the influence range of the action is increased. This phenomena is modeled by Bayes Law.

4 The Agent's Sensing Ability

In this section we discuss how to evaluate the detection function for each agent and explain the concept of *effective volume* for an action of a given agent.

The value of the detection function $\mathbf{b}(a_i, c_j, \mathbf{f})$ for action $\mathbf{f}$ can be obtained by transforming from a pre-recorded standard detection function for a_i and c_j. For a given camera hardware (which is determined by the corresponding agent that uses this camera) and a given recognition algorithm r, the standard detection function $\mathbf{b_0}((\theta, \delta, l), \langle r, w, h \rangle)$ gives a measure of the detecting ability of the recognition algorithm r when no previous action has been applied. In this function, $\langle w, h \rangle$ is the viewing angle size of the camera; (θ, δ, l) is the relative position of the center of the target to the camera:

$$\theta = \arctan\left(\frac{x}{z}\right) \ , \tag{10}$$

$$\delta = \arctan\left(\frac{y}{z}\right) \ , \tag{11}$$

$$l = z \ , \tag{12}$$

where (x, y, z) is the coordinates of the target center in the camera coordinate system. The value of $\mathbf{b_0}((\theta, \delta, l), \langle r, w, h \rangle)$ can be obtained by experiments. We can first put the target at (θ, δ, l) and then perform experiments under various conditions, such as light intensity, background situation, and the relative orientation of the target with respect to the camera center. The final value of b_0 is the total number of successful recognitions divided by the total number of experiments. These values can be stored in a look up table indexed by θ, δ, l and retrieved when needed. Sometimes we may approximate these values by analytic formulas.

We only need to record the detection values of one angle size $\langle w_0, h_0 \rangle$. Those of other sizes can be approximately transformed to those of size $\langle w_0, h_0 \rangle$. Suppose (θ, δ, l) is the target position for angle size $\langle w, h \rangle$, we want to find the value $(\theta_0, \delta_0, l_0)$ for angle size $\langle w_0, h_0 \rangle$ such that

$$\mathbf{b_0}((\theta_0, \delta_0, l_0), \langle r, w_0, h_0 \rangle) \approx \mathbf{b_0}((\theta, \delta, l), \langle r, w, h \rangle) \ . \tag{13}$$

To guarantee this, the images taken with parameter $\langle \theta_0, \delta_0, l_0, w_0, h_0 \rangle$ and $\langle \theta, \delta, l, w, h \rangle$ should be almost same. Thus, the area and the position of the projected target image on the image plane should be almost the same for both images. We obtain

$$l_0 = l \sqrt{\frac{\tan\left(\frac{w}{2}\right) \tan\left(\frac{h}{2}\right)}{\tan\left(\frac{w_0}{2}\right) \tan\left(\frac{h_0}{2}\right)}} \ , \tag{14}$$

$$\theta_0 = \arctan\left[\tan(\theta) \frac{\tan\left(\frac{w_0}{2}\right)}{\tan\left(\frac{w}{2}\right)}\right] \ , \tag{15}$$

$$\delta_0 = \arctan\left[\tan(\delta) \frac{\tan\left(\frac{w_0}{2}\right)}{\tan\left(\frac{w}{2}\right)}\right] \ . \tag{16}$$

When the configurations of two operations are very similar, they might correlate with each other (refer to [17] for detail). Repeated actions are avoided during the search process. When independence is assumed, $\mathbf{b}(a_i, c_j, \mathbf{f})$ is calculated as follows. First, calculate the corresponding (θ, δ, l) of the center of c_j with respect to operation $\mathbf{f}$ of agent a_i. Second, transform (θ, δ, l) into the corresponding $(\theta_0, \delta_0, l_0)$ of angle size $\langle w_0, h_0 \rangle$. Third, retrieve the detection value from the look up table, or get the detection value from a formula.

Now we explain the concept of effective volume $EV(\mathbf{f})$ for a given action $\mathbf{f}$. The ability of the recognition algorithm and the value of the detection function are influenced by the image size of the target. Only when the target can be totally brought into the field of view of the camera and the features detected within a certain precision, can the recognition algorithm be expected to function correctly. So, for an agent's action with a given recognition algorithm and a fixed viewing angle, the probability of successfully recognizing the target is high only when the target is within a certain distance range. We call this range the *effective range* and the viewing volume within this range the *effective volume*. For a given action, only those cubes that are within the *effective volume* can be examined with high detection probability. Figure 4 illustrates in $2D$ the effective volume for a given action.

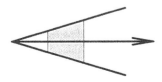

Fig. 4. $2D$ illustration of the effective volume of a given action (the shaded area).

Please note that the influence range $\Omega(\mathbf{f})$ of $\mathbf{f}$ is different from the effective volume $EV(\mathbf{f})$ of $\mathbf{f}$, $\Omega(\mathbf{f}) \neq EV(\mathbf{f})$. The influence range $\Omega(\mathbf{f})$ refers to the region such that if the target center is within this region, the probability of detecting the target is not 0. The effective volume $EV(\mathbf{f})$ refers to the region such that if the target center is within this region, the probability of detecting the target is high enough (higher than a given threshold). So, $EV(\mathbf{f}) \subseteq \Omega(\mathbf{f})$.

5 The Collaborative Multi-agent Search Team: a Global View

A global view of the activities of the multiagent search team is as follows:

- Before the search process, each agent a_i proposes an initial target distribution $(\mathbf{p}(a_i, c, \tau_0\text{-})$ for all $c \in \Omega)$ according to its own perception of the world (Fig. 5).

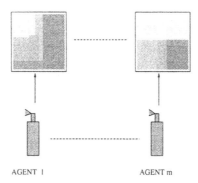

Fig. 5. Each agent has an initial target distribution.

- The team forms the common initial target distribution $(\mathbf{p}(c, \tau_0)$ for all $c \in \Omega)$ by combining the initial distributions of all the agents together (Fig. 6).
- Each agent a_i uses the common initial distribution $\mathbf{p}(c, \tau_0)$ as its own initial distribution $\mathbf{p}(a_i, c, \tau_0)$ used during the search process (Fig. 7). That is, each agent a_i performs the substitution

$$\forall c, \mathbf{p}(a_i, c, \tau_0) \leftarrow \mathbf{p}(c, \tau_0) \ . \tag{17}$$

- The selection and execution of actions for different agents occurs in parallel (Fig. 8). However, action selection and execution for each agent is sequential. This means that a single agent can select and execute the next action only after the current action is finished.
  - Each agent selects its own action based on its own knowledge and the knowledge *learned* from interaction and communication with other agents.

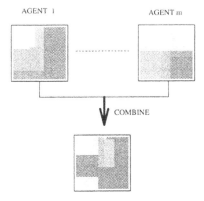

Fig. 6. Common initial target distribution formation.

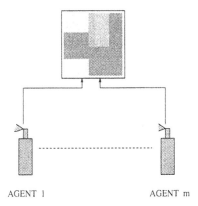

Fig. 7. Initial distribution used before the search process for each agent.

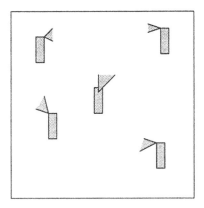

Fig. 8. Agents search for the target together at the same time.

- If the selected action succeeds, the corresponding agent announces the success result immediately and the team search process terminates.
- If the selected action fails, the corresponding agent will broadcast the failure news to the rest of the agents, together with the action's state parameters and effective volume. Then the corresponding agent updates its own knowledge base about the target distribution and starts selecting its next action.

- The team search process will terminate when the available time is used up.

Let $\Omega_{a_i}^{\mathbf{F}}$ be the set of all the actions which can be selected by agent a_i $(1 \leq i \leq m)$. Suppose

$$\mathbf{F}_{a_i} = \left\{ \mathbf{f}_{a_i}^{(1)}, \mathbf{f}_{a_i}^{(2)}, \ldots, \mathbf{f}_{a_i}^{(N_{a_i})} \right\} \tag{18}$$

is the set of actions actually selected by agent a_i $(1 \leq i \leq m)$ from $\Omega_{a_i}^{\mathbf{F}}$ during the search process, where N_{a_i} is the number of actions selected by agent a_i. Then we call $\mathbf{F}_{a_i}$ the *effort allocation* for agent a_i. The *effort allocation for the search team* is the union of the effort allocations of all the agents of the team

$$\mathbf{F} = \mathbf{F}_{a_1} \bigcup \mathbf{F}_{a_2} \bigcup \ldots \bigcup \mathbf{F}_{a_m} . \tag{19}$$

Let $P[\mathbf{F}]$ be the probability of detecting the target by the team effort allocation $\mathbf{F}$ and let K be the total time that can be used by the team. Then the task of object search by a multiagent team can be defined as finding a team effort allocation $\mathbf{F}$ such that $P[\mathbf{F}]$ is maximized within the time constraint K.

5.1 The Initial Common Target Distribution

Before the search process, each agent has some knowledge of the whereabouts of the target. This knowledge can come from different sources. Because the object search team is collaborative, the team should form a common initial target distribution by considering each agent's knowledge. One way to get a common initial distribution is to use the Dempster-Shafer theory [12]. Here we simply generate the initial common target distribution $\mathbf{p}(c, \tau_0)$ (for $c \in \Omega$) by forming a weighted sum of the probability distributions of all the agents. This means that for all $c \in \Omega$, perform the following operation:

$$\mathbf{p}(c, \tau_0) \leftarrow \sum_{k=1}^{m} w_k \mathbf{p}(a_k, c, \tau_{0-}) . \tag{20}$$

The weights $w_1, \ldots, w_m$ must satisfy

$$\sum_{k=1}^{m} w_k = 1 \quad \text{and} \quad w_k \geq 0, \ 1 \leq k \leq m . \tag{21}$$

It is obvious that the newly generated target distribution satisfies

$$\sum_{i=1}^{n,\text{out}} \mathbf{p}(c_i, \tau_0) = \sum_{i=1}^{n,\text{out}} \left(\sum_{k=1}^{m} w_k \mathbf{p}(a_k, c_i, \tau_{0-}) \right)$$

$$= \sum_{k=1}^{m} \left(w_k \sum_{i=1}^{n,\text{out}} \mathbf{p}(a_k, c_i, \tau_{0-}) \right)$$

$$= \sum_{k=1}^{m} w_k$$

$$= 1 \ . \tag{22}$$

In order to make this strategy work effectively, it is important to obtain a set of values $w_1, \ldots, w_m$ which reflect the relative credibility of the target distributions from different agents. If the multiagent team has not been working together before, then the relative reliability is not known and we set

$$w_1 = \ldots = w_m = \frac{1}{m} \ . \tag{23}$$

In general, more reasonable values of w_k can be obtained by training and learning, as follows.

A single training set is obtained by running the team search process for m cases, each case using a different agent's initial probability distribution as the initial global probability distribution (Note: there is no weighted sum operation).

For each case of testing, we can obtain a time needed to detect the target for the search team. Suppose that the time needed to detect the target is minimum for the case when agent k's initial probability distribution is used. Then we say that agent k wins this training set. The result illustrates that agent k's knowledge sources are more reliable than other agent's knowledge sources for this training.

So, for each training set we can select an agent whose initial knowledge is most reliable. Suppose we have performed Q training sets, then the value of w_k for agent k will be the number of trainings won by agent k divided by Q.

5.2 Detection Probability for an Effort Allocation

We derive a general expression to calculate $P[\mathbf{F}]$ in this section in a global view. By global view we mean that we track in a globally consistent fashion the probability of detecting the target by applied actions and update the probability distribution after each action is applied. In other words, if an action can detect the target, then the success news is broadcast at the moment this action is selected, and if an action cannot find the target, then the probability distribution is updated at the moment this action is selected.

In order to get a general expression, let us first consider a special case by assuming that the actions in $\mathbf{F}$ are applied in sequence (Note: in real applications, the actions of different agents can be executed at the same time).

Let

$$F = \{f_1, f_2, \ldots, f_q\} \ , \tag{24}$$

where

$$q = N_{a_1} + \ldots + N_{a_m} = |\ F_{a_1}\ | + |\ F_{a_2}\ | + \ldots + |\ F_{a_m}\ | \tag{25}$$

and

$$\tau_{f_1} < \tau_{f_2} < \ldots < \tau_{f_q} \ . \tag{26}$$

The correspondence of the action index i for f_i with the actions executed by each agent is illustrated in the following

$$F = \left\{ \overbrace{f_1, \ldots, f_{N_{a_1}}}^{F_{a_1}}, \ldots, \overbrace{f_{N_{a_1}+\ldots+N_{a_{m-1}}+1}, \ldots, f_q}^{F_{a_m}} \right\} \ . \tag{27}$$

In other words, $f_{a_i}^{(j)}$, the jth action of agent a_i in its own effort allocation F_{a_i}, is represented as $f_{N_{a_1}+N_{a_2}+\ldots+N_{a_{i-1}}+j}$ in above representation (27).

The search process is as follows: at the beginning, f_1 is selected and applied, the probability distribution used by f_1 is the initial common target distribution. If operation f_i $(1 \leq i \leq q)$ succeeds, then the news is broadcast to all agents and the search process terminates; if operation f_i fails, then the probability distribution is updated, and this updated probability distribution will be used by f_{i+1}. It is clear that the expected probability of detecting the target by allocation F is:

$$P[F] = P(f_1) + [1 - P(f_1)]P(f_2) + \ldots + \left\{ \prod_{i=1}^{q-1} [1 - P(f_i)] \right\} P(f_q) \ . \tag{28}$$

Formula (28) is not helpful for finding regularities about $P[F]$, and it is not appropriate in defining the probability of detecting the target by an effort allocation when the actions of different agents can be executed in parallel. In order to obtain a representation of the detection probability in the general sense, we need to do some algebraic transformations.

The initial probability distribution is denoted as

$$p^{[0]}(c_1), p^{[0]}(c_2), \ldots, p^{[0]}(c_n), p^{[0]}(c_{\text{out}}) \ . \tag{29}$$

Of course, we have $p^{[0]}(c_i) = p(c_i, \tau_0)$ for $i = 1, \ldots, n, \text{out}$. Where $p(c_i, \tau_0)$ $(i = 1, \ldots, n, \text{out})$ are the initial common target distribution formed by integrating the knowledge of all the participating agents before the search process. The probability distributions $p^{[0]}(c_i)$ $(i = 1, \ldots, n, \text{out})$ are used by the first action f_1 of F. After the application of operation f_1, the distributions are denoted by $p^{[1]}(c_1), p^{[1]}(c_2), \ldots, p^{[1]}(c_n), p^{[1]}(c_{\text{out}})$. Generally, after the application of the ith operation f_i, the distributions are denoted by $p^{[i]}(c_1), p^{[i]}(c_2), \ldots, p^{[i]}(c_n)$, $p^{[i]}(c_{\text{out}})$, where $1 \leq i \leq q$. If the target is detected by f_i, then $p^{[i]}(c^*) = 1$ and

$\mathbf{p}^{[i]}(c) = 0$ for $c \neq c^*$, where c^* is the cube that contains the target center. Otherwise, $\mathbf{p}^{[i]}(c_j)$ $(1 \leq j \leq n, \text{out})$ is obtained by using the probability updating rule (9). Let $P(\mathbf{f}_i)$ represent the probability of detecting the target by applying the action $\mathbf{f}_i$ with respect to the effort allocation $\mathbf{F}$. Then of course we have

$$P(\mathbf{f}_i) = \sum_{j=1}^{n} \mathbf{p}^{[i-1]}(c_j)\mathbf{b}(c_j, \mathbf{f}_i) \ . \tag{30}$$

In (30), $\mathbf{b}(c_j, \mathbf{f}_i) = \mathbf{b}(a, c_j, \mathbf{f}_i)$, a is the agent executing action $\mathbf{f}$.

Let $P^{[0]}(\mathbf{f}_i)$ represents the probability of detecting the target by applying the action $\mathbf{f}_i$ when no action has been applied before. This means that we assume $\mathbf{f}_i$ is the first action to be applied, thus the probability distribution used by $\mathbf{f}_i$ is $\mathbf{p}^{[0]}(c_1), \ldots, \mathbf{p}^{[0]}(c_n)$, and $\mathbf{p}^{[0]}(c_{\text{out}})$. We have

$$P^{[0]}(\mathbf{f}_i) = \sum_{j=1}^{n} \mathbf{p}^{[0]}(c_j)\mathbf{b}(c_j, \mathbf{f}_i) \ . \tag{31}$$

After some calculation, we obtain a general expression of the target probability distribution as stated in the following lemma.

Lemma 1. *For allocation* $\mathbf{F}$, *if* $\mathbf{f}_1, \ldots, \mathbf{f}_k$ *have not detected the target, then we have*

$$\mathbf{p}^{[k]}(c) = \frac{\mathbf{p}^{[0]}(c)[1 - \mathbf{b}(c, \mathbf{f}_1)]\ldots[1 - \mathbf{b}(c, \mathbf{f}_k)]}{(1 - P(\mathbf{f}_1))(1 - P(\mathbf{f}_2))\ldots(1 - P(\mathbf{f}_k))} \ . \tag{32}$$

By replacing the result of Lemma 1 and (31) into (30), after some complex transformations we obtain a general expression for the detection probability for any action in $\mathbf{F}$ as follows.

Lemma 2. *For allocation* $\mathbf{F}$, *we have*

$$P(\mathbf{f}_k) = \frac{1}{(1 - P(\mathbf{f}_1))\ldots(1 - P(\mathbf{f}_{k-1}))} \Bigg\{$$

$$P^{[0]}(\mathbf{f}_k)$$

$$+(-1)^1 \sum_{i_1=1}^{k-1} \left(\sum_{c \in \Omega(\mathbf{f}_{i_1}, \mathbf{f}_k)} \mathbf{p}^{[0]}(c)\mathbf{b}(c, \mathbf{f}_{i_1})\mathbf{b}(c, \mathbf{f}_k) \right)$$

$$+(-1)^2 \sum_{1 \leq i_1 < i_2 \leq k-1} \left(\sum_{c \in \Omega(\mathbf{f}_{i_1}, \mathbf{f}_{i_2}, \mathbf{f}_k)} \mathbf{p}^{[0]}(c)\mathbf{b}(c, \mathbf{f}_{i_1})\mathbf{b}(c, \mathbf{f}_{i_2})\mathbf{b}(c, \mathbf{f}_k) \right)$$

$$+(-1)^3 \sum_{1 \leq i_1 < i_2 < i_3 \leq k-1} \left(\sum_{c \in \Omega(\mathbf{f}_{i_1}, \mathbf{f}_{i_2}, \mathbf{f}_{i_3}, \mathbf{f}_k)} \mathbf{p}^{[0]}(c)\mathbf{b}(c, \mathbf{f}_{i_1})\mathbf{b}(c, \mathbf{f}_{i_2})\mathbf{b}(c, \mathbf{f}_{i_3})\mathbf{b}(c, \mathbf{f}_k) \right)$$

$$+$$

$$\vdots$$

$$+(-1)^{k-1} \sum_{c\in\Omega(\mathbf{f}_1\ldots\mathbf{f}_{k-1}\mathbf{f}_k)} \mathbf{p}^{[0]}(c)\mathbf{b}(c,\mathbf{f}_1)\mathbf{b}(c,\mathbf{f}_2)\ldots\mathbf{b}(c,\mathbf{f}_{k-1})\mathbf{b}(c,\mathbf{f}_k)\Bigg\} , \qquad (33)$$

for $k = 2, \ldots, q$.

By replacing the result of Lemma 2 into (28), we can prove the following important lemma.

Lemma 3. *For allocation* $\mathbf{F}$*, we have*

$$P(\mathbf{F}) = \sum_{i=1}^{q}\left(\sum_{j=1}^{n}\mathbf{p}(c_j,\tau_0)\mathbf{b}(c_j,\mathbf{f}_i)\right)$$

$$+(-1)^{2+1}\sum_{1\le i_1<i_2\le q}\left(\sum_{c\in\Omega(\mathbf{f}_{i_1}\mathbf{f}_{i_2})}\mathbf{p}(c,\tau_0)\mathbf{b}(c,\mathbf{f}_{i_1})\mathbf{b}(c,\mathbf{f}_{i_2})\right)$$

$$+$$

$$\vdots$$

$$+(-1)^{r+1}\sum_{1\le i_1<i_2<\ldots<i_r\le q}\left(\sum_{c\in\Omega(\mathbf{f}_{i_1}\mathbf{f}_{i_2}\ldots\mathbf{f}_{i_r})}\mathbf{p}(c,\tau_0)\mathbf{b}(c,\mathbf{f}_{i_1})\mathbf{b}(c,\mathbf{f}_{i_2})\ldots\mathbf{b}(c,\mathbf{f}_{i_r})\right)$$

$$+$$

$$\vdots$$

$$+(-1)^{q+1}\left(\sum_{c\in\Omega(\mathbf{f}_1\mathbf{f}_2\ldots\mathbf{f}_q)}\mathbf{p}(c,\tau_0)\mathbf{b}(c,\mathbf{f}_1)\ldots\mathbf{b}(c,\mathbf{f}_q)\right) . \qquad (34)$$

Lemma 3 explicitly reveals a very important property of $P[\mathbf{F}]$: the order of actions been applied does not have any influence on the value of $P[\mathbf{F}]$. This agrees with common knowledge, because whether an action $\mathbf{f}_{a_i}^{(j)} = \mathbf{f}(a_i, s_{a_i}, r_{a_i}^{(j')})$ can detect the target is determined only by the agent a_i, the state parameters s_{a_i} of a_i, the recognition algorithm $r_{a_i}^{(j')}$ used to analyze the image taken by a_i, the real target position and the real situation (lighting, background, etc.) surrounding the target. Since these factors will not change with time, the time when the action $\mathbf{f}(a_i, s_{a_i}, r_{a_i}^{(j')})$ is applied has no effect on whether this action can detect the target. If action $\mathbf{f}$ detects the target in one application order in the effort allocation $\mathbf{F}$, then, it will detect the target in another application order in the effort allocation $\mathbf{F}$; if action $\mathbf{f}$ fails to detect the target in one application order, then, it will fail to detect the target in another application order.

Since (34) is independent of order of actions, it is used as the general definition of the expected probability of detecting the target by an effort allocation $\mathbf{F}$ in a multiagent search team.

In a real team search process where actions can be applied concurrently by different agents, we will use Formula (34) to calculate the expected probability of detecting the target by the given effort allocation.

In other words, suppose $\mathbf{F} = \mathbf{F}_{a_1} \bigcup \mathbf{F}_{a_2} \bigcup \ldots \bigcup \mathbf{F}_{a_m}$ is the effort allocation of the search team, where $\mathbf{F}_{a_i}$ $(1 \leq i \leq m)$ is defined in (18). Let $\sum_{\mathbf{F}(f_{a_{i_1}}^{(j_1)} \ldots f_{a_{i_k}}^{(j_k)})}$ be the sum operation over all the different subset $\{f_{a_{i_1}}^{(j_1)}, \ldots, f_{a_{i_k}}^{(j_k)}\}$ of $\mathbf{F}$, where $1 \leq k \leq m$. Let $|\mathbf{F}| = N_{a_1} + \ldots + N_{a_m}$ be the number of actions in the team effort allocation $\mathbf{F}$. Then the probability of detecting the target $P[\mathbf{F}]$ by the effort allocation $\mathbf{F}$ is calculated by the following formula:

$$
P(\mathbf{F}) = (-1)^{1+1} \sum_{\mathbf{F}(\mathbf{f}_{a_i}^{(j)})} \left(\sum_{c \in \Omega(\mathbf{f}_{a_i}^{(j)})} \mathbf{p}(c, \tau_0) \mathbf{b}(a_i, c, \mathbf{f}_{a_i}^{(j)}) \right)
$$

$$
+(-1)^{2+1} \sum_{\mathbf{F}(\mathbf{f}_{a_{i_1}}^{(j_1)} \mathbf{f}_{a_{i_2}}^{(j_2)})} \left(\sum_{c \in \Omega(\mathbf{f}_{a_{i_1}}^{(j_1)} \mathbf{f}_{a_{i_2}}^{(j_2)})} \mathbf{p}(c, \tau_0) \mathbf{b}(a_{i_1}, c, \mathbf{f}_{a_{i_1}}^{(j_1)}) \mathbf{b}(a_{i_2}, c, \mathbf{f}_{a_{i_2}}^{(j_2)}) \right)
$$

$$
+
$$

$$
\vdots
$$

$$
+(-1)^{r+1} \sum_{\mathbf{F}(\mathbf{f}_{a_{i_1}}^{(j_1)} \ldots \mathbf{f}_{a_{i_r}}^{(j_r)})} \left(\sum_{c \in \Omega(\mathbf{f}_{a_{i_1}}^{(j_1)} \ldots \mathbf{f}_{a_{i_r}}^{(j_r)})} \mathbf{p}(c, \tau_0) \mathbf{b}(a_{i_1}, c, \mathbf{f}_{a_{i_1}}^{(j_1)}) \ldots \mathbf{b}(a_{i_r}, c, \mathbf{f}_{a_{i_r}}^{(j_r)}) \right)
$$

$$
+
$$

$$
\vdots
$$

$$
+(-1)^{|\mathbf{F}|+1} \left(\sum_{c \in \Omega(\mathbf{f}_{a_1}^{(1)} \ldots \mathbf{f}_{a_m}^{(N_{a_m})})} \mathbf{p}(c, \tau_0) \mathbf{b}(a_1, c, \mathbf{f}_{a_1}^{(1)}) \ldots \mathbf{b}(a_m, c, \mathbf{f}_{a_m}^{(N_{a_m})}) \right) . \quad (35)
$$

We give an intuitive representation of (35). It is clear that $P^{[0]}(\mathbf{f}_{a_i}^{(j)}) = \sum_{c \in \Omega(\mathbf{f}_{a_i}^{(j)})} \mathbf{p}(c, \tau_0) \mathbf{b}(a_i, c, \mathbf{f}_{a_i}^{(j)})$. If we *define* the initial intersect probability of detecting the target by a set of actions $\mathbf{f}_{a_{i_1}}^{(j_1)}, \ldots, \mathbf{f}_{a_{i_r}}^{(j_r)}$ as

$$
P^{[0]}\left(\langle \mathbf{f}_{a_{i_1}}^{(j_1)} \ldots \mathbf{f}_{a_{i_r}}^{(j_r)} \rangle \right) = \sum_{c \in \Omega(\mathbf{f}_{a_{i_1}}^{(j_1)} \ldots \mathbf{f}_{a_{i_r}}^{(j_r)})} \mathbf{p}(c, \tau_0) \mathbf{b}(a_{i_1}, c, \mathbf{f}_{a_{i_1}}^{(j_1)}) \ldots \mathbf{b}(a_{i_r}, c, \mathbf{f}_{a_{i_r}}^{(j_r)}) .
$$

$$
(36)
$$

Then, (35) can be more intuitively written as

$$
P(\mathbf{F}) = \sum_{i,j} P^{[0]}(\mathbf{f}_{a_i}^{(j)}) - \sum_{i_1,j_1,i_2,j_2} P^{[0]}\left(\langle \mathbf{f}_{a_{i_1}}^{(j_1)} \mathbf{f}_{a_{i_2}}^{(j_2)} \rangle \right)
$$

$$
+ \ldots + (-1)^{r+1} \sum_{i_1,j_1,\ldots,i_r,j_r} P^{[0]}\left(\langle \mathbf{f}_{a_{i_1}}^{(j_1)} \ldots \mathbf{f}_{a_{i_r}}^{(j_r)} \rangle \right) + \ldots
$$

$$+(-1)^{|\mathbf{F}|+1} P^{[0]} \left(\langle \mathbf{f}_{a_1}^{(1)} \dots \mathbf{f}_{a_m}^{(N_{a_m})} \rangle \right) \ . \tag{37}$$

5.3 The Task of Multiagent Search and its Complexity: a Global View

The object search task for a multiagent object search team $\{a_1, \dots, a_m\}$ can be defined as the task for each agent a_i to find a set of operations $\mathbf{F}_{a_i}$, such that the expected probability of detecting the target $P[\mathbf{F}]$ calculated by (35) for the effort allocation $\mathbf{F}$ (formed by the union of the effort allocations of all the agents of the team)

$$\mathbf{F} = \mathbf{F}_{a_1} \bigcup \mathbf{F}_{a_2} \bigcup \dots \bigcup \mathbf{F}_{a_m} \tag{38}$$

is maximized with the constraint that the cost for the effort allocation $\mathbf{F}_{a_i}$ for each agent a_i is less than or equal to a given constant K:

$$\begin{cases} \mathbf{t}(\mathbf{f}_{a_1^{(1)}}) + \mathbf{t}(\mathbf{f}_{a_1^{(2)}}) + \dots + \mathbf{t}(\mathbf{f}_{a_1^{(N_{a_1})}}) \leq K \\ \mathbf{t}(\mathbf{f}_{a_2^{(1)}}) + \mathbf{t}(\mathbf{f}_{a_2^{(2)}}) + \dots + \mathbf{t}(\mathbf{f}_{a_2^{(N_{a_2})}}) \leq K \\ \qquad\qquad\qquad\qquad \vdots \\ \mathbf{t}(\mathbf{f}_{a_m^{(1)}}) + \mathbf{t}(\mathbf{f}_{a_m^{(2)}}) + \dots + \mathbf{t}(\mathbf{f}_{a_m^{(N_{a_m})}}) \leq K \end{cases} \tag{39}$$

where $\mathbf{f}_{a_i^{(j)}}$ $(1 \leq j \leq N_{a_i})$ refers to the jth actions of agent a_i.

This is an NP-hard problem. Proof is by restriction. Assume that $m = 1$. The task is to find an effort allocation for agent a_1 such that $P[F]$ is maximized and

$$\mathbf{t}(\mathbf{f}_{a_1^{(1)}}) + \mathbf{t}(\mathbf{f}_{a_1^{(2)}}) + \dots + \mathbf{t}(\mathbf{f}_{a_1^{(N_{a_1})}}) \leq K \ . \tag{40}$$

Since this simplified problem $(m = 1)$ is NP-hard (please refer to [16] for proofs), the multiagent object search problem is also NP-hard. Thus, it is impractical to design a team search strategy that can always generate an effort allocation that maximizes the probability of detecting the target. Thus, the goal becomes for each member of the team to cooperate with the others so as to find the target as early as possible.

6 Learning, Interaction and Planning

The multiagent object search task described in this paper is an ideal task to study the relationships between learning, interaction and planning in multiagent environments.

Ideally, an agent should plan its actions based on a perfect knowledge about the current situation of the world (such as the target distribution), the ability of itself and all the other agents, and the activities of the whole team performed so far. But because of limited computation power, limited memory, and limited communication power, an agent is not able to keep track everything. For example, it is not possible for an agent to update its *own* knowledge base every time *other* agents execute actions that fail to detect the target, because the process of updating the probability distribution is time consuming. In the multiagent search team, the agent updates the probability distribution only after its *own* action is applied.

It is also not efficient if an agent performs the search task only based on its own knowledge without considering other agents actions. This may cause many extraneous activities. Thus, a certain amount of communication is essential for the team search task. The question is how much communication is needed and how much memory an agent must maintain to make the behavior of the whole team satisfactorily coordinated.

An agent's internal knowledge can be divided into two parts: local knowledge and global knowledge. The local knowledge is the agent's knowledge about itself and the influence of its own action on the world. The global knowledge is its knowledge about other agents and the effects of the actions of other agents on the world. Each agent should have a perfect local knowledge, but it is not able to have a perfect global knowledge because of various limitations. Usually the global knowledge is obtained by interaction, communication and learning from other agents during the search process. A search agent's planning system is influenced by both its local knowledge and the learned global knowledge.

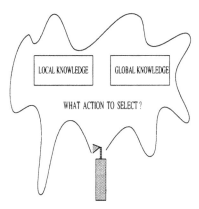

Fig. 9. A search agent's planning system is influenced by both its local knowledge and global knowledge.

6.1 Learning and Where to Look Next

The task of "where to look next" for an agent a_i can be defined as: for a fixed position (x_c, y_c, z_c), select the state parameters (w, h, p, t, r) for the next action $\mathbf{f}$ of agent a_i such that the probability of detecting the target is maximized.

Local Knowledge. If only local knowledge is considered, the task is the same as object search by a single agent, on which we have done extensive research. We have designed a strategy to decompose the huge space of possible sensing actions determined by different (w, h), p, t, and r into a finite set of actions that must be considered and a strategy to select the best action among the resulted actions from the decomposition such that

$$P(\mathbf{f}) = \sum_{c \in \Omega(\mathbf{f})} \mathbf{p}(a_i, c, \tau_{\mathbf{f}}) \mathbf{b}(a_i, c, \mathbf{f}) \tag{41}$$

is maximized [15].

In a multiagent environment, the probability $P(\mathbf{f})$ calculated in this way does not perfectly reflect the real probability of detecting the target by action $\mathbf{f}$, because the knowledge about the target distribution $\mathbf{p}(a_i, c, \tau_{\mathbf{f}})$ used in the calculation is agent a_i's local knowledge, which does not reflect the influences of the actions of other agents (because agent a_i is not able to update the probability distribution whenever other agents apply an action on the environment). So, even though a_i thinks that the region within the effective volume $EV(\mathbf{f})$ of $\mathbf{f}$ has high probability, it might not be so, since other agents may have already examined this region extensively. From the global point of view, the probability within $EV(\mathbf{f})$ may not be so high as a_i thinks at the moment $\tau_{\mathbf{f}}$.

Thus, a_i needs to integrate the influences of other agents' applied actions into its action planning process.

Learned Global Knowledge. Although each agent a_i is not able to update its target distribution whenever another agent executes an action, it is able to record *crude* information about to what extent cells of the region have been checked by other agents.

As discussed before, for a given action $\mathbf{f}$, only those cubes that are within $\mathbf{f}$'s effective volume $EV(\mathbf{f})$ can be checked with high confidence. Thus, whenever an action $\mathbf{f}$ is executed, only the examination situation of those cubes that are within $EV(\mathbf{f})$ need be recorded; the examination situation of other cubes need not be recorded. To do this, each agent a_i maintains the Examination Situation Map $E(a, c)$. $E(a, c)$ gives the number of times that the cell c falls into the effective volume of the actions executed by agent a. Each agent a_i must maintain the Examination Situation Map for every other agent a_j $(1 \leq j \leq m, j \neq i)$ of the search team and every cell c $c \in \Omega$ of the search region Ω (Fig. 10).

For each agent, the Examination Situation Map is maintained through a process of learning by communication. During the search process, as soon as an agent a_i selects an action $\mathbf{f}(a_i, s_{a_i}, r_{a_i}^{(j)})$ to execute, it broadcasts a list of all the cells that are within the effective volume $EV(\mathbf{f})$ of action $\mathbf{f}$ to all the other members of the team. Upon receiving the broadcast from a_i, each other team member a_j $(1 \leq j \leq m, j \neq i)$ updates its Examination Situation Map as following:

$$\forall c \in EV(\mathbf{f}) \text{ perform } E(a_i, c) \leftarrow E(a_i, c) + 1 \ . \tag{42}$$

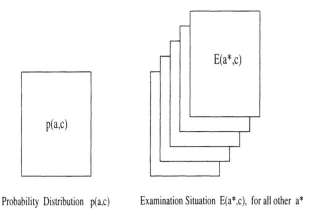

Probability Distribution p(a,c)   Examination Situation E(a*,c), for all other a*

Fig. 10. An agent's local and global knowledge.

The Influence of the Learned Global Knowledge on Planning. The next action $\mathbf{f}$ is selected with reference to both local knowledge and external stimuli. The local knowledge is the probability of detecting the target $P(\mathbf{f})$ by action $\mathbf{f}$. The external stimuli $ES(\mathbf{f})$ is derived from the *learned global knowledge*, which is represented by the Examination Situation Map. The objective function for agent a_i is: select an action $\mathbf{f}$ that maximizes the following weighted sum

$$w_1 P(\mathbf{f}) - w_2 ES(\mathbf{f}) \ , \tag{43}$$

where w_1, w_2 are weights, and $ES(\mathbf{f})$ is defined as follows

$$ES(\mathbf{f}) = \sum_{c \in \Omega(\mathbf{f})} \sum_{j \neq i}^{m} E(a_j, c) \ . \tag{44}$$

6.2 Learning, Coordination and Where to Move Next

The goal of "where to move next" for an agent a_i in a multiagent object search team can be defined as: select the agent position (x_c, y_c) for agent a_i such that the chances of detecting the target for agent a_i and the whole team is maximized.

The Local Knowledge and the Learned Global Knowledge. If only local knowledge is considered, then the task is the same as the "where to move next" task for a single agent. The strategy of selecting the next agent position is straight forward. For each candidate position (x, y), there is a range of space that can be checked by the camera without occlusion. We call this range the sensed sphere $SS(x, y)$ for position (x, y) (please refer to [15] [17] for more detail). The sum of the probability of all the cells within the corresponding sensed sphere is called the sensible probability for this position $S_{prob}(x, y)$. The task is to find a position (x, y) such that the sensible probability is maximized.

As in the previous discussion, the local knowledge $S_{prob}(x, y)$ does not perfectly reflect the real sensible probability for the position (x, y). The agent should also include the learned global knowledge $E_s(x, y)$ into its decision making. $E_s(x, y)$ gives the measurement of how much the region within the sensed sphere of position (x, y) has been checked by other agents

$$E_S(x, y) = \sum_{c \in SS(x,y)} \sum_{j \neq i} E(a_j, c) \ . \tag{45}$$

Coordination. For the "where to move next" task of the multiagent search team, coordination among different agents should also be taken into consideration. For example, if all the agents find the same region to have the highest sensible probability, then when coordination is not considered, all the agents will be gathered around this region to perform actions. This is not desirable for the team as a whole since other regions of the search space are not checked while many of the actions may be redundant. Coordination is needed when deciding the next positions of agents in a multiagent team. From the global point of view, the agents should be distributed evenly across the search space. Thus, a coordination factor $D(x, y)$ should be considered in planning the next action. $D(x, y)$ specifies how far the new position (x, y) is to other agents of the team (Fig. 11).

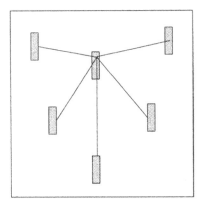

Fig. 11. The relative positions for different agents are also very important for efficient search. The positions of the agents in the team should be carefully selected such that the extraneous activities are avoided.

Planning. The next position (x, y) for agent a_i should be selected by integrating the local knowledge, the learned global knowledge, and the coordination factor together. The goal is to maximize the following weighted term

$$w_1 S_{prob}(x, y) - w_2 E_s(x, y) + w_3 D(x, y) \ , \tag{46}$$

where w_1, w_2, and w_3 are weights, and

$$D(x,y) = \sum_{j=1,j\neq i}^{m} \left[w_{ij}\sqrt{(x-x_j)^2 + (y-y_j)^2} \right] . \tag{47}$$

The value of w_{ij} $(1 \leq j \leq m, j \neq i)$ is used to balance the importance of the distance between agent a_j and agent a_i. These values must satisfy

$$\sum_{j=1,j\neq i}^{m} w_{ij} = 1 . \tag{48}$$

If there is no preference, then we set

$$w_{12} = \ldots = w_{1m} = \frac{1}{m-1} . \tag{49}$$

7 Conclusion

In this paper, we formulate the multiagent object search task and prove that this task is NP-Complete from a global point of view. We analyze various issues that involve learning in a multiagent object search system, including: (A) the local knowledge updating rule for an agent; (B) a method to obtain the initial common target distribution; (C) learning of the global knowledge through interaction and communication with other agents; (D) learning and interaction so as to improve the "where to look next" task; (E) learning and coordination so as to improve the "where to move next" task.

In future work we will endeavor to develop a more detailed interaction and learning theory for the multiagent object search task. Given the highly cooperative nature of this task, we believe that a further study of this problem will reveal deeper relationships between learning, interaction and organizations in multiagent environments.

References

1. Hayes-Roth, B., Brownston, L., Gen, R., V.: Multiagent collaboration in directed improvisation. International Conference on Multi-Agent Systems, U.S.A. (1995)
2. Huber, H., Durfee: On acting together: without communication. AAAI Spring Symposium on Reasoning about Mental States, U.S.A. (1995)
3. Ingrand, F., F., Georgeff, M., P., Rao, A., S.: An architecture for real-time reasoning and system control. IEEE EXPERT. 6 (1992)
4. Jennings, N.: Controlling cooperative problem solving in industrial multi-agent systems using joint intentions. Artificial Intelligence, 75 (1995)
5. Kinny, D., Ljungberg, M., Rao, A., Sonenberg, E., Tidhard, G., Werner, E.: Planned team activity. Artificial Social Systems, Lecture notes in AI 830, C. Castelfranchi and E. Werner, editors. Springer Verlag, New York, U.S.A. (1992)

6. Kitano, H., Asada, M., Kuniyoshi, Y., Noda, I., Osawa, E.: The robot world cup initiative. IJCAI-95 Workshop on Entertainment and AI/Alife, Montreal, Canada (1995)

7. Kuniyoshi, Y., Rougeaux., S., Ishii, M., Kita, N., Sakane, S., Kakikura, M.: Cooperation by observation: the framework and the basic task pattern. IEEE International Conference on Robotics and Automation (1994)

8. Levesque, H., J., Cohen, P., R., Nunes, J.: On acting together. National Conference on Artificial Intelligence, Calif, U.S.A. (1990)

9. Nickerson, S. B., Jenkin, M., Milios, E., Down, B., Jasiobedzki, P., Jepson, A., Terzopoulos, D., Tsotsos, J., Wilkes, D., Bains, N., Tran, K.: ARK: autonomous navigation of a mobile robot in a known environment. Intelligent Autonomous Systems-3, Pennsylvania, USA. (1993) 288–296

10. Pimentel, K., Teixeira, K.: Virtual reality: through the new looking glass. Windcrest/McGraw-Hill, Blue Ridge Summit (1994)

11. Rao, A., S., Lucas, A., Morley, D., Selvestrel, M., Murray, G.: Agent-oriented architecture for air-combat simulation. Technical Note 42, The Australian Artificial Intelligence Institute (1993)

12. Schubert, J.: Cluster-based specification techniques in Dempster-Shafer theory for an evidential intelligence analysis of multiple target tracks. Ph.D Thesis, Department of Numerical Analysis and Computing Science, Royal Institute of Technology, S-100 44 Stockholm, Sweden (1994)

13. Tambe, M., Rosenbloom, P., S.: RESC: An approach for real-time, dynamic agent tracking. Proceedings of the International Joint Conference on Artificial Intelligence, Montreal Canada (1995)

14. Weiss G.: Learning to Coordinate Actions in Multi-Agent Systems. Proceedings of the International Joint Conference on Artificial Intelligence, France (1993) 311–316

15. Ye, Y., Tsotsos, J.: Where to look next in 3D object search. 1995 IEEE International Symposium for Computer Vision, Florida, U.S.A. (1995) 539–544

16. Ye, Y., Tsotsos, J.: Sensor planning in 3D object search: its formulation and complexity. The 4th International Symposium on Artificial Intelligence and Mathematics, Florida, U.S.A. (1996)

17. Ye, Y., Tsotsos, J.: Sensor Planning for Object Search. Technical Report, RBCV-TR-94-47, Computer Science Department, University of Toronto. (1994)

Evolution of Coordination as a Metaphor for Learning in Multi-Agent Systems

Ana L. C. Bazzan[*]
Forschungszentrum Informatik
University of Karlsruhe
Haid-und-Neu-str. 10-14
76131 - Karlsruhe, Germany
e-mail: bazan@fzi.de

Abstract. In societies of individually-motivated agents where the communication costs are prohibitive, there should be other mechanisms to allow them to coordinate. For instance, by choosing an equilibrium point. Such points have the property that, once all agents choose them, none can get higher utility from this joint decision. Hence, if this fact is common knowledge among agents, then an introspective reasoning process leads them to coordinate with low communication requirements. Game theory offers a mathematical formalism for modelling such interactions among agents. In classical game theory, agents or players are assumed to be always rational, which is a strong assumption regarding bounded-rational agents. Moreover, agents do not profit from results of past interactions. In the evolutionary approach to game theory on the other hand, agents do not need to have full knowledge about the rules of the game. Instead, they are involved in a learning process through active experimentation in which they may reach an equilibrium by playing repeatedly with neighbors. Such evolutionary approach is used here to model the interactions in a network of autonomous agents in the domain of control of traffic signals. The game is one of pure coordination with incomplete information. The dynamics of the game is determined by stochastic events which affect the traffic patterns forcing agents to adapt to them by altering their current strategies. Due to such perturbations, the equilibrium of the system eventually changes as well as the behavior of agents. If the game lasts long enough, agents can asymptotically learn how to re-coordinate their strategies and reach the global goal.

Keywords. distributed artificial intelligence, multi-agents systems (MAS), learning in MAS, coordination in MAS, game-theoretical approach.

1 Introduction

One of the central research topics in multi-agent systems is how to coordinate individually-motivated agents. This kind of agents usually have their own intentions,

---

[*] the author is supported by Conselho Nacional de Pesquisa Científica e Tecnológica - CNPq - Brasil.

plans and knowledge, and are willing to solve their own local goals. If the global goal of the system is to solve a given problem, then a coordination mechanism is necessary to settle the conflicts which may arise either due to the need of allocating limited resources, or due to agents having opposite intentions. Former coordination mechanisms were mostly based on communication, as for instance the contract-net protocol introduced by Smith and Davis (1981). In fact, the communication bottleneck is a major shortcoming in their framework.

The goal of our research is to develop an approach which minimizes, and possibly eliminates the need of communication when coordinating the agents' actions. In some domains like, for instance traffic signals control, the communication cost is high. Moreover, a communication-based coordination is very slow.

By bringing independent and individually-motivated agents to coordinate, it is desirable that they reach a state of coordination which is not only stable, but also as favorable as possible for every participant. Some researches which have been carried out in this direction have an economic flavor (as for instance Rosenschein and Zlotkin, 1994 and Wellman, 1992). We shall restrict ourselves to the game-theoretical approach proposed by the former. Indeed game theory provides tools for modelling the interaction among agents as well as their deductive process. Such tools lead them to anticipate the opponent's action and select the best answer to it. Conflicts are embedded in the utility functions of the players and solved implicitly as agents select the equilibrium points of the game.

In classical game theory this deductive or introspective process is possible because it is assumed that players are rational and have common knowledge to some extent. These are of course strong assumptions since players are not always fully rational in their actions. Bounded-rational players do experiment because they prefer to sacrifice the immediate gain in order to learn more about the rules of the interaction they are involved in. Apart from introspection, an alternative explanation as to how players choose their actions is to assume that they are able to extrapolate from what they have observed in past interactions, provided they have been involved in similar games. In this way the equilibrium point can be learned, which seems a more plausible assumption. Therefore, players do not need to know how their actions influence their opponents' actions. Eventually, they learn that opponents do not play certain strategies, thus replicating the iterative dominance solution concept of classical game theory. Even if players know only their own payoffs, they may eventually converge to a steady state. As each one learns its opponents' preferences, the system asymptotically converges to a state represented by a set of evolutionary stable strategies (ESS), which are equilibrium points.

This approach was first discussed by Maynard Smith and Price (1973) regarding evolution of populations of animals genetically programmed to play different strategies. As the more successful have higher fitness and as natural selection favors those which are fittest, after some generations the population reaches an ESS and cannot be invaded by a mutant strategy. In their work they were concerned with the stability of equilibria against a mutation or a perturbation in the environment. After this work the interest in such evolutionary techniques have also motivated studies related to economics and political science, where the main concern has been the behavior of systems on the long run, as well as learning the optimal one-shot action by means of playing repeatedly.

We introduce this analysis to the problem of coordinating individually-motivated agents acting in the domain of traffic signal control. The aim is to synchronize traffic signals positioned at neighboring intersections on a road. Agents are involved in a long run interaction in order to select signal plans which allow them to coordinate towards a specific strategy, thus permitting the traffic on the whole road to flow in a better way. The main motivation for this approach is that it has the benefit of coping with interactions in which agents are fully individually-motivated and need only to know their own payoffs and not those of the opponents, hence modelling coordination with low or no communication requirements. Another positive characteristic is that a complex n-player game can be modelled as several 2-player games.

Our emphasis is towards learning amidst bounded-rational agents in dynamic systems. Learning here means experimentation and adaptation as in reinforcement learning. We assume that agents do not know anything regarding the game and learn about it from their past observations. In game theory, this topic is explored in Kalai and Lehrer (1993), although in their research they assume that players start with subjective beliefs about individual strategies used by each opponent. We do not make this assumption, thus following the research lines of Sastry *et al.* (1994) and Kandori *et al.* (1993). The former consider a game of incomplete and imperfect information in which players do not even need to know that opponents exist. After some periods of interaction, players update their probability distribution concerning each of the strategies, based solely on the current action and the payoff received. As players have no information concerning others, they learn in a decentralized fashion.

Learning is a consequence of *individual* payoff maximization plans and is achieved as the game progresses. In this sense it may be seen as *learning by playing* (Kalai and Lehrer, 1993) in games of incomplete and imperfect information. There is a criticism that this kind of passive learning can lead to steady states which are not equilibria. Need of active learning, i. e. with experimentation, was emphasized in Fudenberg and Levine (1993) who claim that even if the players play the same game many times, they may continue to hold incorrect beliefs about the opponents' preferences unless they perform enough experimentation. In our approach, this is done through genetic operators.

In the next section we give a brief overview on solution concepts from both classical and evolutionary game theory regarding the situations where rationality fails. In Section 3 a learning algorithm called "EVO Algorithm" is presented. In Section 4 we discuss our implementation of the evolutionary approach in the traffic signal control scenario. In Section 5 a measurement of performance is presented, as well as the results of our simulations. We then conclude with Section 6.

2 Evolution as a metaphor for learning a coordination point

The classical game-theoretical modelling assumes that the players are rational, i.e. they all make decisions in pursuit of their objectives, namely the maximization of the expected payoff in a utility scale. Game theory offers mathematical tools for modelling conflicts which arise in antagonistic interactions among agents which have individual, possible conflicting goals. Thus, game theory serves as a basis for agents to predict a coordination point which is the equilibrium point of a game representing an interaction. This has been the motivation for the use of game-theoretical techniques in multi-agent systems as a coordination mechanism (e.g. Rosenschein, 1985). Coordination is achieved by solving the game through introspection, i.e. by finding the best way of playing the game for every player. The best known solution concept is the Nash equilibrium given by a set of strategies, one for each player, where each one is the best response to the strategies of others. It is the one that yields the greatest payoff to each and guarantees at least one equilibrium point, eventually in mixed strategies. A set of strategies S is a Nash equilibrium iff:

$$\forall i, \, p_i\left(s_i^*, s_{-i}^*\right) \geq p_i\left(s_i', s_{-i}^*\right), \, \forall s_i' \quad \text{where } \pi \text{ is the payoff function and } s_{-i}^* \text{ indicates}$$

the strategies of all other players.

The problem with the Nash concept is that the equilibrium point may not be unique. However, uniqueness is crucial for the majority of multi-agent interactions modelled by using game theoretical techniques. Agents which rely on this modelling to negotiate with other agents may not be able to come to an agreement if they are not sure of which action other agents will choose. Therefore a refinement process is required which rules out some equilibria, eventually leading to a unique equilibrium point.

Although some refinements of the Nash concept exist, game theory has been unsuccessful in explaining how players know, for instance, that a Nash equilibrium will be played. Moreover, the classical game theory fails to explain how players choose one Nash equilibrium if a game has multiple equally plausible equilibrium points. Introspective theories that attempt to explain equilibrium at the individual decision-making level by means of rationality impose very strong informational assumptions and are widely recognized as having serious deficiencies. Assuming that each player knows all about the structure of the interaction in which he is involved, this knowledge may not be enough for him to decide how to play, for he must also predict the move of his opponents. The understanding of this process requires an explanation of how players' predictions are formed. One classic explanation is that an outside mediator suggests the strategy profile to each player, who accepts it unless he could gain by deviating. A second classic explanation is that the game is solvable by dominance and it is common knowledge that players are Bayesian rational, so that player's introspection leads them all to the same predictions. A more recent theory to explain how players anticipate a solution is to assume that they are able to learn from what they have observed in past interactions, provide they have played similar games.

The dynamics of evolutionary games draws strongly from the biological literature in which natural selection mechanisms favors the fittest animals of a population (i.e. those whose strategy yields a higher reproductive fitness or payoff against the population). This approach was first discussed by Maynard Smith and Price (1973) regarding evolution of populations of animals genetically programmed to play different (pure) strategies. As natural selection favors those which are more successful, after some generations the population may reach an evolutionary stable equilibrium (ESS), which is an equilibrium point having the property that it cannot be invaded by any mutant strategy. In their work they were concerned with the stability of equilibria against mutations represented by one-shot perturbations.

After Maynard Smith and Price, the interest has shifted also to economics and political science, where the main concerns have been on the long-run behavior of the system, and on learning the optimal one-shot action by means of playing repeatedly. Experiments with the repeated Prisoner's Dilemma (Axelrod, 1984) represent an important milestone regarding evolution as a metaphor for learning. Nowak and May (1992) also tackled this problem and first explores the pattern which appears on a two-dimensional $n \times n$ grid. In each site of the grid a player plays the repeated game only with its nearest neighbors. Their pure deterministic approach, in which players have no memory, was later criticized in Huberman and Glance (1993) who have run an asynchronous updating of strategies, claiming that in this case the coexistence of cooperators and defectors was not possible any longer.

Kandori *et al.* (1993) on the other hand, have tackled coordination games, in which there exists more than one strict Nash equilibrium. Which one to select has been the concern of many researches in classical game theory, notably from Harsanyi and Selten (1988) as to what regards their risk dominance solution concept. Kandori *et al.* (1993) showed that there is a unique stationary distribution on the number of players playing each of the strategies, which is achieved independently of the initial choice of individual strategies. In games with multiple strict Nash equilibria, some are more likely to emerge than others when in the presence of continuously small stochastic shocks.

The coordination mechanism proposed here aims at allowing agents to learn from past interactions, instead of relying on pure introspection. Such mechanism is based on an experimental learning process. Even if agents possess only little local knowledge, which they get from sensoring their near environment, they are able to perform experimentation and, according also to the decisions taken elsewhere in the neighborhood, they receive a reinforcement in their payment function. This feedback can be either communicated to each agent, or be locally detected. In this paper we are concerned with the case where there is no communication, hence agents are involved in a pure self-organization process where they receive a reinforcement due to their actions performed in the near past and this value is obtained from their local sensors only.

Additionally, we use techniques related to genetic algorithms to code in strings the strategies chosen in the recent past, i.e. during the time interval between two learning processes. The stochastic events which may take place in the network and which affect the strategies to be chosen, are modelled by mutations. During the learning process, a fitness for each string is computed and it influences the next generation of strategies which will be used by the agents to perform the experimentation. Depending on the frequency of the stochastic events, the agents are able to coordinate towards the strategy which corresponds to the global goal.

When considering a game having a unique iteratively undominated strategy s and played dynamically, on the long run only s will be played. A plausible explanation is that players learn how to play their best replies, once the learning process requires strategies that are not doing well to be played less often.

Evolution as a metaphor for learning in societies of bounded-rational players has been the subject of several studies like those cited above. General assumptions are a set of random matching in a population large enough, and myopic or bounded behavior of players. Myopia means that an action which was learned as being effective remains effective in the future. The myopic behavior can be explained because players may have high discount rates compared to the speed of adjustment of the system. Another assumption which is often made is inertia, justified by the cost of changing strategy. Some shortcomings arising from all the above assumptions are:

- bounded-rational players ignore that opponents are also engaged in similar learning processes, therefore ignoring that what is being learned keeps changing; inconsistencies may occur and players may not learn or may converge to false beliefs;
- players may not perform experimentation.

Different approaches have been described to tackle these shortcomings, all of which account for distinct degrees of bounded rationality, informational and communication requirements (which decreases from the first to the last approach below):

- all players know their payoff matrix, which specifies the payoffs for *every combination of actions* taken by the opponents. Also they have initial subjective beliefs about opponents' strategies and are informed of all previous actions taken (Kalai and Lehrer, 1993);
- a small portion of players is informed about the distribution of strategies in the population. As the rest, once they are uninformed, do not change their strategy, those informed can react by playing their best response to that distribution (Fudenberg and Kreps, 1989);
- players are not able to calculate a best response and just use other players' best strategy as entry for some kind of automaton which gives them a response as output. In this category lies most of the research related to economical aspects and also that of Nowak and May (1992);
- players neither observe the distribution of strategies in the population nor are able to calculate best responses. However interaction are repeated and players have enough time to observe the behavior of their opponents. Besides, there is a rate of experimentation due to mutant players. These players select a strategy at random, are informed about the payoff, and attempt to learn the necessary reaction to the opponents' strategies.

While in biological applications of evolutionary game theory the ESS is genetically determined, in more general cases players can learn such strategies. This can be done by analysing the payoff got from each rule used to select strategies in the past. According to Harley (1981), in a set of actions $A_i = (a_1,...,a_m)$, a learning rule is a rule which specifies the probabilities $P = \left(P_{i,1,t},...,P_{i,m,t}\right)$ as a function of the payoffs obtained by playing those strategies in the past. He has also defined a rule for learning an ESS as a rule which causes the members of a population with any initial $P = \left(P_{i,1,0},...,P_{i,m,0}\right)$ to adopt the ESS of the game after a given time. He proved that such a rule must have the following property:

$$P(a_k) > \frac{\text{total payoff for playing } a_k}{\text{total payof so far}}.$$

Therefore, in the near future, each strategy is selected according to its probability. This leads the ESS to be selected asymptotically. It is assumed that the payoffs correspond to changes in fitness, and that the game is played enough times to ensure that the payoffs received after an ESS is reached exceed the payoffs received during the learning period.

However, this kind of passive learning can lead to steady states which are not equilibria. Fudenberg and Levine (1993) claim that even if players play the same game many times, they may continue to hold incorrect beliefs about the opponents' preferences unless they perform enough experimentation. This can be done in several ways, such as considering mutations or players sometimes selecting strategies at random. In the model proposed by Harley, on the other hand, learning rules lead to the ESS without completely fixing or deleting any possible strategy. In fact, in his model a poor strategy a_k is never completely discarded because it may become advantageous if the environment changes. This is done by weighing the recent payoffs more than the older ones, without however forgetting the whole past.

3 The EVO Algorithm

In our model, agents are involved in a pure self-organization process where they receive a reinforcement (also called feedback or payoff) due to their actions performed in the near past, and this value is acquired from their near environment. This is to be verified in a network of individually-motivated agents. A network K is represented by a finite graph in a two-dimensional lattice space with no empty sites allowed. A site is occupied by one agent selecting mixed strategies. A connection between two sites represents the interaction between agents occupying these sites.

Each individually-motivated agent positioned at neighboring site on the network has local goals which determine its actions. Agents are involved in a long run interaction in order to select an action which address their local states. However, as they are paid also according to a global goal, agents have an incentive to coordinate towards this goal.

Each agent i plays a game G against each member of his neighborhood N_i. G is represented by the 3n-tuple $(1,...,n, A_1,...,A_n, \pi_1,..., \pi_n)$ where:

$I = (1,...,n)$ is the set of players or agents, where player n is the Nature (often, a generic player i represents all elements of K, that means excluding n);

$A_i = (a_{i,1},...,a_{i,k},...,a_{i,m})$ is the set of pure strategies of i;

the mapping $\pi_i: \underset{j \in I}{X} A_j \to \Re$ is the payoff function of i;

$\mathbf{P_i} = \left(P_{i,1},...,P_{i,k},...,P_{i,m}\right)$ (the mixed strategy of i) is a probability distribution on A_i;

$P_{i,k}$ is the probability assigned to the k-th pure strategy of i, with $P_{i,k} \geq 0$ and $\underset{k}{\sum} P_{i,k} = 1$ for each $k \in A_i$;

S_i is the set of all mixed strategies of i;

$S = \underset{i \in I}{X} S_i$ is the mixed strategy combination of G.

Each $i \in I$ updates its mixed strategy based solely on the payoff received by selecting an action, irrespective of other agents' moves and payments. Therefore, the game is of incomplete and imperfect information. $\mathbf{P_i}$, the mixed strategy for i, is time varying. After selecting an action $a_{i,t} \in A_i$ at time t, each i in K receives an individual payoff, calculated as the sum of the payoffs obtained by playing G against each $j \in N_i$.

The dynamics of the model is as the algorithm called "EVO algorithm" for short:

```
begin;
t := 0;
set initial values of the parameters (pi, pl, pg, r, K, Pi's, Q, T, λ);
repeat for each i in K while not last period;
    t := t + 1;
    repeat while not a global-state-change;
        pool sensor and determine a change in the local state;
        repeat while not an individual-state-change period;
            repeat while not a learning period;
                perform action ai,t;
                collect payoff πi,t;
                accumulate payoff πi,t;
            compute new fitness Fi,t and probability vector Pi,t due to learning;
        compute new probability vector Pi,t due to a change in local state;
end;
```

Before the actual beginning of the game, the types of the agents and their payoff function are set by the Nature. Her selection of the payment represents the stochastic events which occur at each site. Nature determines the payoff function for each possible combination of mixed strategy s in S. In each subsequent period, agents select a mixed strategy with a probability which is determined by their beliefs about their environment. These beliefs are formed by gathering the information regarding the environment by means of a sensor, and by learning the efficiency of the actions selected in the past.

A period in which $\mathbf{P_i}$ changes, no matter in which site i of K, either by learning or by a local stochastic event, is called a *learning period* (for i) or an *individual-state-change period* (for i) respectively. Others are normal *payoff-getting periods*.

3.1 Individual-state-change Periods

The local stochastic events happen with probability p_i at each site in an independent way. Local change events are detected at each site by pooling the sensors. The information acquired at the k-th detector $(q_{i,k})$ is compared with each $q_{i,l}$, for each k,l belonging to the set of sensors of i. Each agent i then classifies the detectors in decreasing order by the flow of vehicles and keeps this information stored in a stack.

Each time there is a change in the top position of the stack, an event at a local level (for i) is said to happen. Upon such an event, a change in the distribution of the mixed strategy is required.

When a local event occurs at time $t=\rho$ at site i, agent i updates the vector $\mathbf{P_i}$ as a function of the value $q_{i,k}$ measured at each corresponding k-th sensor:

$$\mathbf{P_{i,t}} = (P_{i,1,t},\ldots,P_{i,k,t},\ldots,P_{i,m,t}) = [d(q_{i,1,t}),\ldots,d(q_{i,k,t}),\ldots,d(q_{i,m,t})] \text{ with}$$

$$d(q_{i,k,t}) = \frac{q_{i,k,t}}{\sum_k q_{i,k,t}} \quad \text{and for each } k \in A_i : \ P_{i,k,t} \geq 0 \ ; \ \sum_k P_{i,k,t} = 1$$

$$\tag{1}$$

At $t=0$, Equation 1 is used to set the initial vector $\mathbf{P_i}$. This distribution is then employed in the selection of strategies in the subsequent periods, until a learning period or an individual-state-change period occurs and it is updated.

3.2 Payoff-getting Periods

The global payment function is known by the Nature but remains unknown to the agents. These know only their local states by means of the sensors. It is assumed that Nature is informed about the global changes in the network, which can be detected in determined points of it. Therefore, Nature is able to set the correspondent payment functions for the agents according to the vector $\mathbf{P_n}$ on A_n. For example, Figure 1 schematically shows such a move for a 2x2 coordination game. For instance, if the network condition requires agents to select an action s_1, agents actually selecting this action are paid better. Nature changes the payment functions when a global alteration in the network state occurs.

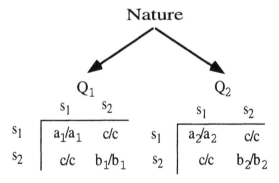

Fig. 1: Payoff Matrices Q for both moves of Nature in a 2x2 Coordination Game

Each individual payment thus depends on the payoff matrix selected by Nature and on the actions selected by the agents. If they are being paid by Q_1 (where $a_1 > b_1$, $c < a_1$, $c < b_1$), and both agents select the action, say s_1 of $A_i = (s_1, s_2)$, then they both receive a payoff of a_1. If they select s_2 they receive a payoff of b_1. Otherwise they receive a payoff of c.

Both configurations $E_1 = (s_1, s_1)$ and $E_2 = (s_2, s_2)$ are Nash equilibria in pure strategies. A third equilibrium, in mixed strategies, is reached when strategy s_1 is played with probability $b_1/_{a_1+b_1}$ and s_2 is played with probability $a_1/_{a_1+b_1}$, in case the entire population is being paid by the payoff matrix Q_1. In the opposite case, the equilibrium is probability $b_2/_{a_2+b_2}$ on s_2 and probability $a_2/_{a_2+b_2}$ on s_1. This equilibrium is undesired since the whole population is expected to coordinate towards the same strategy. It is expected that, with time, only the pareto-superior equilibrium be selected, i.e. (s_1, s_1) if Nature is paying agents according to Q_1, or (s_2, s_2) if Nature is paying them according to Q_2.

At each payoff-getting period, i plays a two-player coordination game with each of the elements $j \in N_i$, and selects a mixed strategy $\mathbf{P_{i,t}} = (P_{i,1}, ..., P_{i,k}, ..., P_{i,m})$ on A_i. The "raw" payoff for agent i when the joint strategy $s = (a_i, a_j)$ of actions is selected at time t is $\pi_{i,j,t}^{o}(s)$. Player i receives a summation of payoffs from these games. Hence, the payoff received by i at time t is given by:

$$\pi_{i,t}^{+}(s) = \sum_j \pi_{i,j,t}^{o}(s) \qquad j \in N_i \qquad (2)$$

Let $\tau > 0$ be a time interval (generally τ represents the time between the last learning period and the current period), $a_{i,t} \in A_i$ be the action agent i selects at time t, and $a_{i,k}$ each k-th pure strategy available to selection. The payoffs received in the last τ periods are represented by the vector $\pi_{i,\tau}^{*}$:

$$\pi_{i,\tau}^{*} = (\pi_{i,1,\tau}^{*}, ..., \pi_{i,k,\tau}^{*}, ..., \pi_{i,m,\tau}^{*}) \qquad (3)$$

where $1 \le k \le m$ is a pure strategy $a_{i,k} \in A_i$.

Finally, each element of the vector $\pi_{i,\tau}^{*}$ at $t \in \tau$, $t \ge 1$ is computed as following:

$$\pi_{i,k,t}^{*} = \begin{cases} \pi_{i,k,t-1}^{*} + \pi_{i,k,t}^{+} & \text{if } a_{i,k} = a_{i,t} \\ \pi_{i,k,t-1}^{*} & \text{otherwise} \end{cases} \qquad (4)$$

3.3 Learning Periods

If participants of a multi-agent system are to reduce or avoid the communication required for a near-optimal coordination process, learning capabilities are even more important in this circumstance than in the standard AI scenario. In the model considered here, each agent has to show the capability of adapting to the stochastic changes in the environment (other agents included), and improving its performance.

The emphasis here is towards learning amidst bounded-rational agents in a dynamic environment. It is assumed that agents do not know anything regarding the strategies and actions of the neighbors, and deduce it from their past observations. The process of adaptation to the stochastic events is done by means of natural evolution techniques, like genetic transformations on the strategies played by the population over time. A performance function based on the feedback (payoffs) obtained from the environment is used as fitness. Such feedback enables the agent to "learn from its mistakes". Depending on the frequency of the stochastic events, agents have time to learn rules about how to change strategies and are able to coordinate towards the global goal.

In order for the pure strategy a_k with the highest expected value of environmental reaction to be selected, agents must learn using a selection rule. Moreover, it is required that agents' selection rules do not assign zero probability to any pure strategy a_k. In the EVO algorithm, the learning rule assigns greater significance on recent than on past payoff information. To achieve this, a memory factor λ ($0 < \lambda \leq 1$) is used in order to avoid the complete neglect of the payoffs obtained by one action in the past time. At each period, the more recent payoff yielded by a given action is reduced by a factor of ($1-\lambda$) as shown in Equation 5.

The learning process does not occur at each period of time. Learning periods happen randomly and are determined by a learning frequency parameter f_l. At each period, there is a probability p_l for each agent to learn. If $t=\theta$ is a learning period for i and $t=\rho$ ($\rho<\theta$) is the last period in which an individual-state-change has happened before θ, then Δ is the learning interval, i.e. the time interval between the θ and ρ, for each θ. The "reduced" payoff, i.e. the payoff which also accounts for $\overline{\pi}_{i,k,\Delta}$ (the average payoff yield by the action $a_{i,k}$ during the interval Δ), then reads:

$$\pi_{i,k,t} = \lambda \, \pi_{i,k,t}^* + (1-\lambda) \overline{\pi}_{i,k,\Delta} \text{ for each } k \in A_i \tag{5}$$

The cumulative and the average payoff after τ yield by action $a_{i,k}$ can be calculated by:

$$\pi_{i,k,\tau} = \sum_{\theta=t-\tau}^{t} \pi_{i,k,\theta} \tag{6}$$

and:

$$\overline{\pi}_{i,k,\tau} = \frac{\pi_{i,k,\tau}}{\tau} \tag{7}$$

Equations 6 and 7 are also used to compute $\pi_{i,k,\Delta}$ and $\overline{\pi}_{i,k,\Delta}$ respectively, when $\tau=\Delta$ is the time interval as defined above.

The learning process consists of agents updating their $\mathbf{P_i}$ vectors according to the efficiency of strategy in the past. This is a function of the fitness vector $\mathbf{F_i}$ defined over the elements of the set of strategies A_i and is computed locally as a function of $\pi_{i,k,\Delta}$:

$$\mathbf{P_i} = \mathbf{F_i} = (F_{i,l,\Delta},\dots,F_{i,k,\Delta},\dots F_{i,m,\Delta}) =$$

$$= \left(\frac{\pi_{i,l,\Delta}}{\sum_k \pi_{i,k,\Delta}},\dots, \frac{\pi_{i,k,\Delta}}{\sum_k \pi_{i,k,\Delta}},\dots \frac{\pi_{i,m,\Delta}}{\sum_k \pi_{i,k,\Delta}} \right) \quad l \le k \le m \; ; \; k \in A_i \tag{8}$$

This probability distribution on the pure strategies is then used in the subsequent periods until a new learning period or a change in the local environment happens. In time, coordination may emerge as the limit behavior of the population.

4 Traffic Signal Coordination Scenario

One scenario we have simulated is a network K consisting of a set of agents $I = \{1,\dots,n\}$ assigned to intersections (Fig. 2). The goal in the traffic signal coordination is to bring as many neighbors as possible to use the same strategy (here signal plan) since these are designed to allow vehicles to flow through the intersections in the main directions, namely eastward (E) or westward (W), without stopping at red lights. A coordination towards E or W is desired, depending on which one of these directions shows the higher flow of vehicles. In this sense, a neighborhood for i is defined as N_i and is composed of the agents immediately related to agent i, which are located along the E and W direction only, since typically the heavy traffic flows in these direction and the side streets in the N-S direction play a minor role for the control of the network.

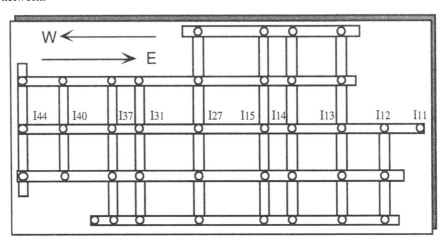

Fig. 2. Network of Traffic Intersections

Figure 2 shows such a network. N_i here is $\{i_l,i,i_r\}$, i.e. agent i and the left and right neighbors. Strategies played by the upper and lower neighbors do not matter for the decision process of i because we are considering only the main roads in the E and W directions. Hence the traffic signals aligned along those directions should coordinate towards one of them in order to deal with the flow of vehicles. Figure 3 depicts an arterial in which the main directions are either west (W) or east (E). Each agent i has a set of actions $A_i = \{sp_W, sp_E\}$ to choose among before actually running the signal plan. The fact that agents give priority to one direction only means that this direction is allotted more green time, but of course the other directions also receive green, and the general constraints posed by safety rules like minimum green time for each lane, minimum and maximum cycle time, etc. were respected when designing the signal plans.

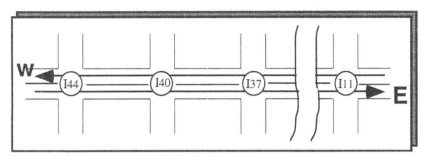

Fig. 3: Direction of synchronization (E or W) in an arterial.

Besides, each intersection has local information coming from sensors (here traffic detectors) on the main lanes which deliver data about the traffic state on these lanes. Having this information, an agent i is able to decide which strategy to use. If currently signal plan sp_W, is running, hence giving priority to direction W, but direction E is demanding more time than W, the agent has to put higher probability on selecting signal plan sp_E. This change certainly modifies traffic states at the neighborhood to some extent, thus eventually leading the immediate neighbors to also change to sp_E. In any case, i must learn which strategy has been more profitable.

As to what concerns the use of evolutionary techniques in the traffic signal coordination scenario, our research differs on a few points from those cited in the previous sections. First, in the traffic scenario the state of the network lasts hardly more than a few hours. After a major change happens, agents should re-coordinate and reach an equilibrium in short time. Of course it depends on how the thresholds that demand a change in traffic signal plans are set. Generally speaking the equilibrium, once reached, will last until a major change involving several neighbors happens.

Once an agent perceives a local change in its traffic state and changes strategy, it does not have reasons for believing that the neighbors will continue doing what they have done so far, because they all will get different payoffs from those they got in the past. In this way, it behaves like a new agent with no further knowledge apart from the local one and have to readapt to the new situation. Depending on the number of agents in this condition and of course whether they interact among themselves (are in the same neighborhood), the equilibrium point will eventually change. In general, for one agent to coordinate towards the opposite strategy it is necessary that it be surrounded by neighbors already running this strategy most of the time.

The second particularity in the traffic scenario is a consequence of its relatively unstable nature: it is desirable that agents do not need to communicate each change to the neighbors because this is not cost-effective. We thus depart from the standard assumptions made in classical game theory by not requiring that the agents have full knowledge of each others' strategies and that they have known prior distributions on any parameters of the game.

Finally another important difference is that in the traffic scenario, it makes no sense for each agent to interact with the entire population. The interaction is reduced to the neighborhood N_i. It is worth noting that in a small population an equilibrium is reached faster but can still be upset by low change rates.

At each period, agents interact separately with the neighbors in a 2-player game and can play one of the strategies as an answer to traffic states measured by the detectors at the main lanes of the intersection. According to the traffic state, it is better either to play sp_W, or sp_E. In each case, the choice results in a payoff which also depends on the strategy chosen by the neighbor. Due to the lack of communication among agents, the general traffic pattern remains unknown to them. If the trend is that traffic flows predominantly in the direction W, for instance, agents running strategy sp_W, are better paid than those running strategy sp_E.

All stochastic events are addressed at discrete periods of time t=1,2,... . Figure 4 shows an example of a sequence of such events. As already discussed, at the beginning of the game the traffic flow pattern and other parameters are defined by a move of Nature. In the example discussed below, agents are paid according to the values shown in the matrix Q_1 of Fig.1. This information is not explicitly communicated to them, but they are able to deduce it by reading the detectors values. Then based on its local knowledge only, each agent has to decide which strategy to play at each period of time. The choices are recorded in a string. Depending on the strategies played in the neighborhood, each agent gets a payoff which is summed up regarding each strategy played.

At each period, there is a probability p_l for each agent to learn how good the set of strategies played in the near past was by computing the fitness of its string. The more fit a string of strategies, the more the agent considers it a good response to the behavior of the neighbors. Therefore, the agent runs it proportionally more often, until it detects a major change in the traffic situation at the intersection, i.e. a change which requires a change in the strategies played. Until such change is detected, the agent is better off if running the set of strategies which proved to be the best in the past. By doing this, agents may reach a coordination of traffic signals which is the global goal.

Besides the learning probability, at each stage agents have also a small probability of experiencing a local change in traffic state at a local level. By allowing agents to react to local changes, the equilibrium may shift because one equilibrium point can upset the other if a sufficient number of agents suffers such local changes. The higher the rate of these changes, the more unstable the system. However there is another parameter which affects stability: the population size. For sufficiently large populations, an equilibrium can only upset another after an extremely long time. As populations here are small, this is not likely to happen.

A change in strategy selected at site i can either invade the neighborhood if it causes a modification in neighbors' traffic states or die out. In this last case, it dies also on the site where it first happened, because after n periods the agent i learns to coordinate again and abandons its mutant strategy. This shows how the global goal of a neighborhood, namely coordination, upsets a local goal.

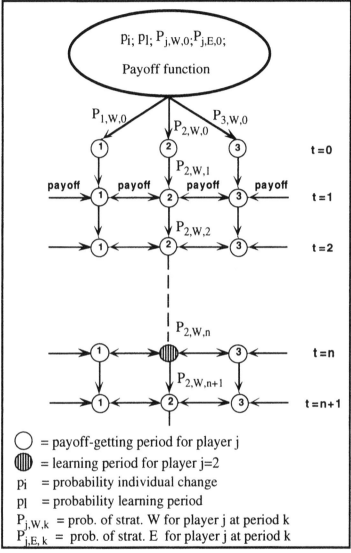

Fig. 4: Learning Process for one agent and two neighbors

5. Experiments

5.1 Measurement of Performance

All simulations discussed next assume $I = 10$, $A_i = \{ sp\,W, sp_E \}$. In order to model the events which happen in a real traffic network, a tool was developed in which the strategies chosen by the agents within time can be simulated using the EVO algorithm. Initially, the average payoff received by the agents were recorded against time. However, such average payoff is only a poor a measure of the global performance of the network during the simulation time τ since it says little about *which agent* miscoordinated, for how long, and in which site the miscoordination happened since this average value can be reached by several different configurations.

Another measurement was tried, namely the variation of the quantity of each type of interactions occurring between all pairs of neighbors. These can be of three types: WW (both selecting sp_W,), EE (both selecting sp_E), and miscoordinated (WE+EW) (i.e. they select either sp_W, sp_E or sp_E, sp_W). Once the total of each type is recorded, they can be cumulatively plotted against time. In fact, the best measure of the stability of an equilibrium is the behavior of the interactions of miscoordinated type since the number of EE stabilizes more quickly. This happens because for an agent selecting E in a population where neighbors select W most of the time, the probability of one of them selecting E (thus making an EE), is lower than the probability of selecting W, thus making an miscoordination. Such plotting give a qualitative measure of the number of agents reaching coordination. However, it says little about time. When should one consider that coordination on WW or EE is reached? The assumption that agents continually maximize their expected payoffs is quite strong for infinitely repeated games. While this assumption is common in game theory and economics, the solution of such a maximization problem in infinitely repeated games may be very demanding. For instance it is not realistic to expect that *all* pairs of agents coordinate *all* the time, since this would mean that the local conditions are not addressed when they are in conflict with the global goal.

A much more interesting measurement of the performance is the evolution of the average probability agents put on selecting the action which is better paid (determined by Nature). One expects that all agents put increasingly higher probability on the more profitable action, and that on the long run they select this action with probability close to one. How close these curves are, depends on both the memory factor λ, and on the frequency of individual changes in traffic state, since this requires that one or more agents respond to these changes in a random way.

As the experiments were repeated several times, the behavior of agents from repetition to repetition of the simulation may vary significantly. However, with time, nearly all behaviors asymptotically tend to the selection of the action which gives more green time to the direction W (sp_W). This is depicted in Figure 5, in which average curves for several values of the parameters p_l and p_i are plotted.

While these mean curves qualitatively characterizes the performance of agents, there is still a need to evaluate these curves concerning the time needed to reach a given pattern of convergence. This is done by reading the time needed to reach the probability $P_{n,W}$ in the curve, that means, the probability Nature puts on paying agents better for the more profitable action (in this case sp_W). Since all simulations depicted in Figure 5 were carried out setting $P_{n,W}=0.9$, agents are expected to be able to put at least an average probability of $P_{i,W} = P_{n,W} = 0.9$ on their selection of sp_W, in order for the situation simulated to be considered a good one.

5.2 Results

Several scenarios have been simulated, varying the learning and the individual-state-change rates and letting agents choose their strategies. At each period, Nature pays *all agents* according to her knowledge of the global traffic condition, represented by the vector $\mathbf{P_n}$. Hence, in these experiments, it is assumed that agents are paid by matrix Q_W ($a=2$, $b=1$, $c=0$) with probability $P_{n,W}=0.9$, and by matrix Q_E ($a=1$, $b=2$, $c=0$) with probability $1-P_{n,W}=0.1$. The immediate interest here is to find the lowest f_i (frequency of individual change) which still guarantees that agents learn how to coordinate and reach the global goal. This frequency is a function of the probability p_i: at each period there is one chance in f_i for an individual change in traffic condition at each intersection. Simulations are thus done setting f_i to 10, 20, 50, 100, 200, and 300 and the learning frequency f_l to 5 and 10.

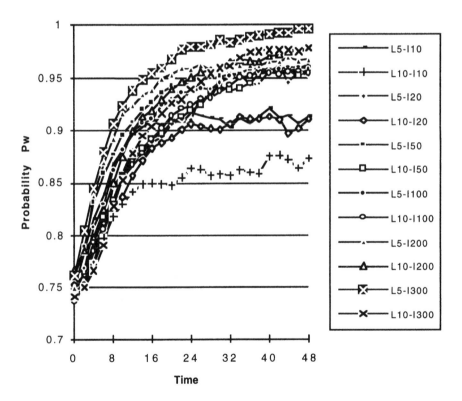

Fig. 5: Evolution of the probability of selecting the global-efficient strategy sp_W against time. Comparison of mean curves for λ=0.95.

Mean curves for each of these conditions are shown together in Figure 5. The time (t_n) needed for the population of agents to reach $P_{i,w}=0.9$ is measured for each condition and depicted in Table 1. As expected, a good pattern of coordination is reached faster in the most stable environment, i.e. with f_i=300. However, the learning rate plays also a role, once, in general, it takes less time for a population with f_l=5 and lower f_i (e.g. f_i =200) to reach this pattern than for a population with higher f_i (e.g. f_i =300) and f_l=10. If learning periods occur frequently, the neighborhood is quickly led to the equilibrium in mixed strategy, when a portion of the agents play one equilibrium and another plays the other equilibrium. This is of course undesirable. On the other hand, if the learning frequency is low, it takes too long for agents to reach an equilibrium. In our scenario, it is desirable that agents react to a change and learn how to coordinate in a new situation within few cycles of the signal plan, i.e. within minutes.

In order to analyse the influence of the memory factor, similar experiments were done with λ=0.80. Analysing the results, one sees that the memory factor also has an effect on the time needed to reach a given pattern of coordination. In the case in which the population is expected to perform poorly (t_n=22 for λ=0.95, f_l=10, and f_i=20), by setting λ=0.80 the population even fails to reach $P_{i,w}$=0.9 within the simulation time (50 periods). For an intermediate situation with f_l=10 and f_i=100, t_n does not differ for the three memory factors. And finally, in the situation expected to be the best of those compared, i.e. in a stable environment, results were the same. Here there is thus room for employing an even lower memory factor if necessary.

Table 1: Time (tn) needed to reach P$_w$=0.9 for curves depicted in Figure 5 (λ=0.95).

$f_l\,(p_l)$	$f_i\,(p_i)$	time (t_n) (periods)
5 (0.2)	10 (0.1)	14
	20 (0.05)	19
	50 (0.02)	12
	100 (0.01)	12
	200 (0.005)	9
	300 (0.003)	8
10 (0.1)	10 (0.1)	x
	20 (0.05)	22
	50 (0.02)	19
	100 (0.01)	18
	200 (0.005)	13
	300 (0.003)	10

6 Conclusion

Although game theory has played an increasing role as a mathematical tool and language in multi-agent systems, its assumption concerning rationality is widely recognized as too strong. Moreover, in biological and physical systems, it is questionable whether, for instance animals or traffic signals behave rationally. Another important criticism is that the classical game theory does not attempt to profit from the fact that agents can learn a deal by playing repeatedly. Both of these concerns are tackled by evolutionary game theory. Rationality is replaced by natural selection and utility payoff by fitness (as a measure of how efficient a strategy is).

Therefore, evolution has increasingly been seen as a metaphor for agents to learn how to reach an equilibrium in a game.

Concerning the state-of-the-art in multi-agents systems, these can profit from such an evolutionary learning in several ways. For instance in chapter 6 of Rosenschein and Zlotkin (1994) the authors discuss scenarios where agents do not know the value of their actions but do know their goals. In our approach, agents are involved in an adaptation process where they learn their goals by active experimentation. For example in the scenario discussed here, agents perceive only their *local* goals which are not necessarily the major goal in the neighborhood.

The approach discussed here proved to be efficient for the traffic signal control domain since in this scenario the traffic situation changes often and a coordination based on a explicit negotiation framework would demand an amount of communication which is not affordable in real systems. Moreover the approach is flexible enough to allow that, once an equilibrium is reached, agents are not stuck in it if Nature has introduced changes in traffic patterns. By adapting, they are able to reach the coordination state which is appropriated for the new traffic situation.

As for the extensions of the approach, it would be desirable to investigate how communication of intentions between neighbors affect their future actions, specially if an agent receives two communications demanding him to act in opposite directions. Agents have to find out, through the reinforcements received, which neighbor has been trustworthy, which one has contributed better, and so on.

Acknowledgements

The author would like to thank Prof. Dr. U. Rembold from the University of Karlsruhe for his advice and support, and the referees for their comments and suggestions. This work has been carried out at the Forschungszentrum Informatik (FZI), Dept. Technical Expert Systems and Robotics.

References

Axelrod, R. (1984). *The Evolution of Cooperation*. Basic Books, New York.

Fudenberg, D. and D. Kreps (1989). A Theory of Learning, Experimentation and Equilibrium in Games. Stanford University. Cited in Kalai and Lehrer, 1993.

Fudenberg, D. and D. Levine(1993). Steady State Learning and Nash Equilibrium. *Econometrica*, **61**: 547-573.

Harley, C. B. (1981). Learning the Evolutionary Stable Strategy. *Journal of Theoretical Biology*, **89,** 611-631.

Harsanyi, J.C. and R. Selten (1988). *A General Theory of Equilibrium Selection in Games*. Cambridge, Mass., MIT Press.

Huberman, B.A. and N. S. Glance (1993). Evolutionary games and computer simulations. *Proc. Natl. Acad. Sci.*, **90**: 7716-7718.

Kalai, E. and E. Lehrer (1993). Rational Learning leads to Nash Equilibrium. *Econometrica*, **61**: 1019-1045.

Kandori, M., G.G. Mailath and R. Rob. Learning, mutation and long run equilibria in games. *Econometrica*, **61**: 29-56.

Maynard Smith, J. and G. R. Price (1973). The logic of animal conflict. *Nature*, **246**: 15-18.

Nowak, M.A. and R.M. May (1992). Evolutionary games and spatial chaos. *Nature*, **359**: 826-829.

Rosenschein, J.S. and M.R. Genesereth (1985). Deals among Rational Agents. In: *Proc. of the Int. Joint Conf. on Art. Intelligence.*

Rosenschein, J.S. and G. Zlotkin (1994). *Rules of Encounter.* The MIT Press, Cambridge (MA)-London.

Sastry, P.S., V.V. Phansalkar and M.A.L. Thathachar (1994). Decentralized Learning of Nash Equilibria in Multi-Person Stochastic Games with Incomplete Information. *IEEE Trans. on Systems, Man and Cybernetics,* **24**, 769-777.

Smith, R.G. and R. Davis (1981). Frameworks for cooperation in distributed problem solving. *IEEE Trans. on Systems, Man and Cybernetics*, **11**, 61-70.

Wellman, M. P. (1992). A general equilibrium approach to distributed transportation planning. In *Proc. of the tenth National Conf. on Artificial Intelligence*, San Jose, California.

Correlating Internal Parameters and External Performance: Learning Soccer Agents

Rajani Nadella & Sandip Sen

Department of Mathematical & Computer Sciences

University of Tulsa

rajani@euler.mcs.utulsa.edu, sandip@kolkata.mcs.utulsa.edu

Abstract. We have developed a Soccer simulator in Java, which allows us to evaluate agent behavioral strategies for effectively collaborating with other team members while countering threats posed by the opposing team. Whereas action selection mechanisms to decide when to shoot, pass, dribble, guard opponent, tackle, etc. are important from a player's viewpoint, learning the capabilities of individual opponents from repeated encounters can provide critical information for the success of a team. Each agent is described by a set of skill levels for different soccer skills. A soccer agent uses its incomplete perceptions, model of the changing environment, knowledge of the skill levels of its own team players, and estimated skill levels of the opponent team players to select the most prudent action. In this paper, we identify learning opportunities for soccer agents, and investigate some of these possibilities in detail. Initial experimental results demonstrate the advantage of learning agents.

1 Introduction

In order to perform controlled experimentation to evaluate the role of different individual and group biases, strategies, protocols, etc., well-designed, artificial environments can be used as effective testbeds [5]. We are building such an environment in which synthetic agents cooperate and compete in environments which we can control and alter to observe effects of environmental variations on the appropriateness of agent control strategies. The domain under investigation is a robotic soccer domain (though it should more appropriately be called a "softbot"-ic soccer domain because only simulated software, and not hardware-based, agents are playing the game). Recently, a couple of papers have reported developments using different robotic soccer simulators [9, 12]. Whereas these papers concentrate on skill learning to improve the capability of a player, our work is complementary in the sense that we investigate behavioral strategy choices (both at individual and group level) while assuming given skill levels of the players. Though our choice of the robotic software domain was motivated in part by the Robot World Cup Initiative [8, 10], we have decided to build our own soccer simulator because our goal is to use this simulator as a testbed for experimentation of individual and group behavioral strategies and to study several key, often closely related, issues in the design of agent systems. These include responsiveness-predictability tradeoff, agent modeling, learning and adaptation,

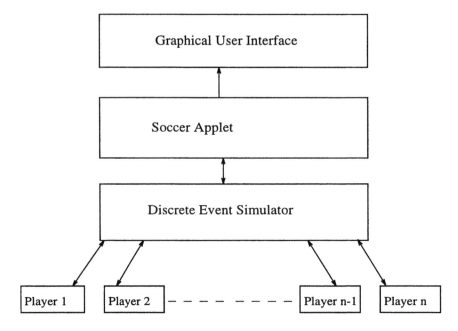

Fig. 1. Components of soccer simulator.

reasoning with limited information, arriving at a consensus, and dealing with environmental uncertainty. Our work is also different from other recent papers that concentrate on applying organizational design principles for designing teams of soccer agents [3].

In a complex domain like soccer, where agents need to cooperate with teammates and compete with opponent team players at the same time, agents must intelligently choose their actions based on the models of other players. Whereas agents may have a reasonably good models of their own teammates (at least the about the static components of agent characteristics), they need to build opponent models from experience and then use these models to make rational decisions. In this paper, we identify the scopes for several learning scenarios by which agents can improve their models of other agents, deal with environmental uncertainty and take calculated risks in the presence of environmental uncertainty. Though all the examples used in this paper is taken from the simulated soccer domain, we believe the learning mechanisms and scenarios proposed are generalizable to other domains.

2 Simulated soccer

The components of our soccer simulator, and their interactions, are depicted in Figure 1. The soccer simulator is a discrete event simulator [2] developed in Java [4]. The discrete event simulator module can simulate any of a set of predetermined actions. It is also responsible for maintaining the world state and updating it periodically. The soccer applet module includes processes required to create and update the graphical interface of this system. The effects of the actions chosen by the player are depicted in the graphical interface. The player module contains the defining characteristics of a player (like shooting efficiency, tackling efficiency etc.), models of other players, and a set of behavioral strategies to determine appropriate actions.

An object-oriented design philosophy is adopted for the design of the simulator. Each player in the player module, for example, is defined as an object with a given set of skills and behavioral strategies. In our current implementation, the simulator instantiates twelve players, six each for two teams. Each team consists of a goalkeeper, left and right fullbacks, one midfielder, and left and right forwards.

The set of player skills consist of the following parameters: speed with ball, speed without ball, dribbling efficiency, shooting efficiency, passing efficiency, throw-in efficiency, tackling efficiency. The goalkeepers have an extra parameter, goalkeeping efficiency, that determines their goalkeeping effectiveness. Each parameter value is selected in the range $[0, 1]$. We set these values for the players such that the sum total of the parameters are the same for each player. Forwards have better skills corresponding to attacking play (e.g., dribbling efficiency, shooting efficiency, etc.) but lesser skills corresponding to defensive play (e.g., tackling efficiency).

At any point in time the position of a player is given by its location (x and y coordinates) and an orientation (an angle $0 \leq \theta \leq \pi$) which specifies the direction in which the player is looking. Players have a constant angle of vision, α, centered at their orientation. So, a player can only see other players and the ball if they fall within its cone of vision. The set of players and ball locations that a player can see at any point in time is referred to as its *viewpoint*.

After each time tick, the simulator calculates the viewpoint of each of the players. Then, for each player, the simulator transmits the corresponding viewpoint and requests an action. The set of actions available to players are the following:

run: this command is used to change location by players who do not have possession of the ball;

dribble: this command can be used by the player in possession of the ball if it wishes to continue running with the ball;

pass: this command can be used by the player in possession of the ball if it wishes to pass the ball to another location;

shoot: this command can be used by the player in possession of the ball if it wishes to shoot at the goal;

tackle: this command is used by a player to attempt to tackle the opponent player with the possession of the ball if they are within a pre-defined distance of each other;

throw-in: when a pass from a player crosses a side-line of the field, the opponent team is awarded possession of the ball, and has to use this command to bring the ball into play again;

assume-location: the players can issue this command to move to arbitrary positions on the field in the next time tick only when "dead-ball" situations occur. These situations include goal-kicks, throw-ins, and free-kicks (the only occasion for free-kicks in our current implementation occurs when a player is caught off-side; fouls and other infringements are not included in our simulation – our players are all completely 'fair'!).

Each player chooses an action depending upon the status of the game, its position, and the information it has learned about the opponents. Player positions and their models are updated after each time tick of the simulator. The simulator is also responsible for resolving any conflicts (e.g., one player trying to pass when an opponent is trying to tackle it), and for enforcing the rules of the game by ignoring invalid commands (e.g., tackling from too far).

3 Learning opportunities

Some of the desirable capabilities of autonomous agents include monitoring the performance of themselves and behavior of other agents in the system as well as sharing information with other agents as and when necessary. The soccer simulator allows us to explore and evaluate a number of strategies concerning various dimensions of agent behaviors.

In order to make effective action selections, agents need to have a real understanding of their own capabilities as well as that of the other players on the field (including teammates and opponents). Though each agent knows its own skill levels as well as the skill levels of its teammates, they can understand the implications of these numbers only by observing how their actions translates into performance on the "soccer field". For example, a passing efficiency of 0.7 is supposed to lead to more accurate passing than a passing efficiency of 0.5, but the actual passing errors observed when pass attempts are made are the only feedbacks each agent can use to find out the passing accuracy implications of its passing efficiency skill level.

In the soccer domain, there exist several possibilities for an agent to improve its understanding of the world and the models of opponent players. In the following we describe a few of the useful individual and group learning scenarios in the soccer domain:

- An agent learns the correlation of its passing efficiency and the passing accuracy that this translates into in the simulated world by repeatedly observing the deviation of an actual pass from the intended target point. We assume that the agents know the factors affecting the accuracy of the pass (passing efficiency of the passer and the length of the pass) and can ignore possible extraneous factor (e.g., things like wind speed, etc. that can affect a pass in real life are not accounted for in our simulation).
- An agent can learn when to shoot and when not to shoot at the goal. By analyzing a history of shooting events (successes and failures from different ranges), a player can develop a good estimate of its effective shooting range.

- By repeatedly observing or engaging an opponent in a tackling situation an agent can arrive at an estimate of its tackling or dribbling skills.
- Once agents develop approximate models of opponents, they might find it prudent to re-assign team roles. We are also investigating the negotiation strategies that agents use to decide whether to switch assignments. First they must combine individual estimates of the skill levels of the opponent team members to arrive at a consensus rating. We plan to use a weighted voting scheme that takes into consideration the amount of past observations used by each player for their ratings [13]. That is, individual estimates are weighted by the number of times they have updated their models to arrive at the current ratings. Then a switch is made if the new match-ups is estimated to be more effective for the team than the current match-up. Thus individually acquired knowledge by team members can be synthesized to produce strategic realignment of the team. Here we find a classic case of individual learning providing the information needed for group learning.

An interesting aspect of some of the learning scenarios described above is the fact that an agent can simultaneously learn the correlation of its skill level, environmental uncertainty, and the skill levels of other players. For example, whether a pass is successful or not depends on skill levels of the passer, the recipient, opponent players in the vicinity and the stochastic variation in the environment that affects where the ball will land.

Other researchers are investigating competitive and cooperative learning opportunities in robotic software domain [1, 9, 12]. Most of this work uses well-known machine learning techniques like reinforcement learning schemes [7] and back-propagation algorithm [11]. These learning procedures are computationally intensive in nature and require sizable amounts of data to learn accurate state-action mappings. We are interested in procedures that can provide reasonable performance accuracy with significantly less computational effort and limited training data. In the following we will identify some inductive inference procedures that may be less sophisticated than the above-mentioned techniques but serve adequately well as means for developing models of an agent's own skills and that of other agents.

3.1 Learning to make effective passing decisions

In this section, we explain the mechanism we employ to enable a soccer agent to learn to make effective passing decisions. When an agent wants to pass a ball to its teammate the decision whether or not to pass the ball is based on the following data: own passing skill level, the length of the pass, the skill levels and locations of the opponent players guarding the teammate to which the ball has to be passed.

We first recount, in more detail, the mechanism used by the simulator to carry out a pass command. When a player decides to pass a ball to a point (x, y) at a distance d, two circles are drawn whose radii is directly proportional on d and inversely proportional to p_{pe}, the passing efficiency of the defender. The

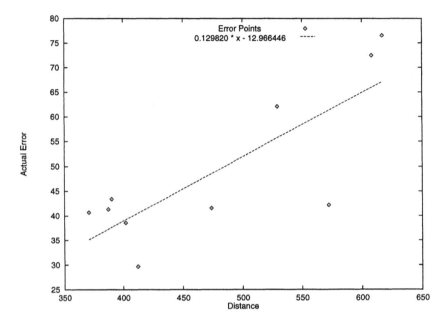

Fig. 2. Using linear regression to calculate expected error from observed errors.

radius of the two circles are given by $r_{low} = \frac{c_1 * d}{p_{pe}}$ and $r_{high} = \frac{c_2 * d}{p_{pe}}$, where $c_1 < c_2$ (hence $r_{low} < r_{high}$). Two concentric circles are drawn with (x, y) as center and with radii r_{low} and r_{high}. The ball is then placed randomly anywhere outside the smaller circle but inside the larger circle. This procedure is meant to simulate a passing error that is bounded above and below and dependent on the length of the pass and the passing skill of the passer. Once the ball location is calculated as above, the player who can get it to quickest is given the ball.

The purpose of the learning scheme in this scenario would be to predict the deviation of error between the desired location and the actual location in which the ball lands. We are particularly interested in online learning mechanisms that are computationally cheap and work with few data points. We believe that though this learning mechanisms are not as accurate as off-line time and data intensive learning mechanisms like neural networks, genetic algorithms, etc., it can provide agents with greater flexibility in changing environments. As such, we formulated the learning to predict passing error problem as the problem of learning to approximate a stochastic function. A computationally cheap mechanism that provides a coarse fit to sample data points (each point in this problem is a pair comprising of the distance of the pass attempted and the amount of error) is the linear regression scheme which fits a straight line through the scattered points so as to minimize the sum squared errors [6]. If (x_i, y_i) represent the ith of n data points, then the equation for the linear regression straight line is given

	Equal	Better	Worse	Variable
Learning	90 (51.2)	93 (77.6)	62 (2.6)	87 (70.5)
Always	75 (100)	88 (100)	40 (100)	76 (100)
Random	73 (50.2)	89 (51.5)	43 (53.5)	75 (46.7)

Table 1. Table of successful pass percentages for different passing decision mechanisms under different match-ups (numbers in parentheses denote the percentage of time a pass was decided).

by $y = mx + c$, where

$$m = \frac{\alpha}{\beta}, \quad \beta = \sum_{i=1}^{n}(x_i - \bar{x})^2, \quad \alpha = \sum_{i=1}^{n} y_i(x_i - \bar{x}),$$

and $c = \bar{y} - m\bar{x}$, where $\bar{x}$ and $\bar{x}$ are the average x and y values respectively.

An agent accumulates data points from its past passing actions. It calculates the linear regression line when it gets two or more data points. After it has accumulated sufficient number of data points, it continues to calculate the regression over the k most recent data points (we have used $k = 10$). This allows agents to track any changes in the dynamics of the environment. Figure 2 presents the regression line constructed by a player from its past passing experiences.

The decision mechanism for passing is as follows. First it decides where to pass the ball (for now, we have assumed that it always passes a fixed distance in front of the teammate); let the target point be (x_t, y_t). The distance to the target point and the passer's passing efficiency is used to find an expected error, ϵ, as predicted by the function obtained from the learning mechanism described above. Next, the nearest opponent to the teammate is located. Assume that the location of the teammate and the nearest opponent are (x_p, y_p) and (x_o, y_o) respectively. Next the worst estimated location of the ball is calculated as the point (x_e, y_e) which is at a distance ϵ from (x_t, y_t) along the straight line joining (x_t, y_t) and (x_o, y_o). Now, if based on the known running speed of the teammate and estimated running speed of the opponent it is found that the teammate will get to (x_e, y_e) from (x_p, y_p) quicker than the opponent can get to it from (x_o, y_o), the decision is taken to pass the ball. Else, the ball is not passed. If the ball is passed exactly to the teammate (i.e., $(x_t, y_t) = (x_p, y_p)$), then the following theorem holds:

Theorem: *If the learning mechanism does not underestimate the passing error, the above decision mechanism guarantees a successful pass.*

But, our learning mechanism does not provide the required guarantee (it is fairly simple to design one that does, but the corresponding error estimates are likely to be large enough to lead to a "not pass" decision in most circumstances, which is counter-productive). In order to experimentally evaluate the proposed learning scheme, we ran experiments with the passer placed at the

(0,0) coordinate of the field and creating 1000 passing situations. Each situation placed one teammate randomly in the rectangle whose diagonally opposing corner points are (200,0) and (600,400). Two opponents were also placed within a short distance of the teammate such that the angle between the two opponents with respect to the teammate is at least 90 degrees. A pass is simulated in every situation as described above. The passing success rate is incremented or decremented depending on whether the teammate or an opponent gets the ball respectively.

We evaluated the effectiveness of our learning scheme against two other decision mechanisms: one that always passes, and one that passes randomly half the time. We experimented with four different player skill sets, each skill set comprising of the running efficiency of the teammate and the two opponents. The passing accuracy used is 0.8. The four skill sets represent different match-ups of skill levels and are labeled as follows:

Equal: teammate - 0.8, opponent 1 - 0.8, opponent 2 - 0.8;
Better: teammate - 0.8, opponent 1 - 0.5, opponent 2 - 0.5;
Worse: teammate - 0.5, opponent 1 - 0.8, opponent 2 - 0.8;
Variable: teammate - 0.8, opponent 1 - 0.6, opponent 2 - 1.0;

The performance of the three decision mechanisms on the four match-ups are presented in Table 1. The data in the table clearly shows that learning based decision mechanism outperformed the others under all situations. Another interesting observation is that in the match-up with all players having equal speed (**Equal**), although the learning mechanism and the random passing mechanism chose to pass the ball almost the same number of times, the successful pass completion rate of the learning mechanism is much higher. This clearly shows the effectiveness of the learning mechanism.

3.2   Estimating opponent skill level

To be effective, a soccer agent need not only learn about its own potential, but at the same time develop a model about the skill levels of players of the opponent team. Using these models, a team can make necessary adjustments to either exploit weaknesses in the opponent team or counter threats posed by the opponent team.

In the following, we present some initial results from experiments where a soccer agent learns about its opponents skill level from experience. When a defender and a forward from opposite teams engage in a tackle, both can estimate the skill level of the opponent given its own skill level and the outcome of the interaction. For example, if the defender's tackling efficiency is te and it wins the tackle with a particular forward, then it can infer that probably the dribbling efficiency of the forward, de is less than te. However, since the defender knows that the outcome is probabilistic, it is not certain of this inference. By repeatedly interacting with the same forward, the defender may be able to obtain a better estimate of that forward's dribbling efficiency. Now, suppose after N interactions,

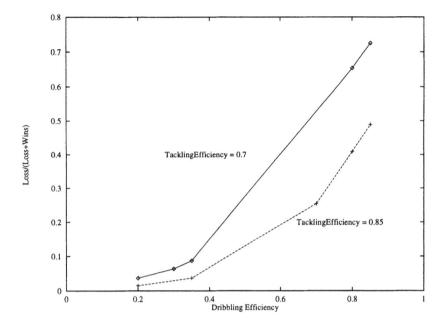

Fig. 3. Observed tackling failure ratios of two defenders against their teammates.

the defender sees that it lost δ% of the tackles. How is it going to estimate de? What the defender needs is a function to map from tackling failure percentage to the opponent's dribbling efficiency. An approximation to such a function can be built from the defender's experience of engaging its own teammates "during practice."

The learning during practice is an off-line learning opportunity where a player can repeatedly engage in tackling situations with other members of the same team. From these interactions, a player can get a set of points describing what its expected tackling failure (measured as a percentage of tackles) is when playing against players of known dribbling efficiencies. Such points are shown in Figure 3 for two defenders with tackling efficiencies of 0.7 and 0.85. To use these off-line experiences to estimate opponent forward's dribbling efficiency, a piecewise linear function f is constructed by each defender from the set of points in its experience. Then, the inverse of this function, f^{-1}, and observed tackling failure data with a given opponent is used by the defender to estimate that opponent's dribbling efficiency.

We also observed that due to the probabilistic nature of the outcome of the tackles, initially there is are wild swings of the estimate of the opponent dribbling efficiency. For example, if the first tackle is lost and the second is won, then the observed loss ratio changes from 1 to 0.5 which leads to a large difference in the estimated dribbling efficiency. To somewhat counter this rapid observed fluctuations, we decided to average the estimate of team members for the same opponent. Figure 4 the individual estimates of two defenders, as well as

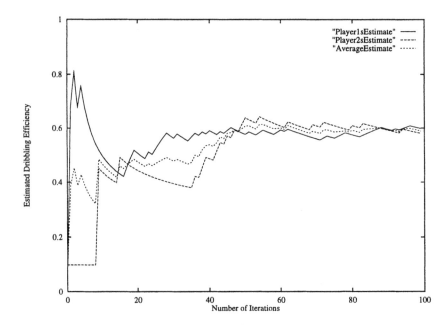

Fig. 4. Estimates of the dribbling efficiency of an opponent forward by two defenders over repeated tackling experiences.

their combined estimate, of the dribbling efficiency of an opponent over different interactions. From the figure, it is clear that the combined estimate has much less fluctuations and hence shared learning can be quite effective in the soccer domain. After sufficient number of interactions the estimated efficiency is found to be very close to the actual value of 0.6.

The accuracy of the estimate depends on the distribution of dribbling efficiencies of the team members of the learning agent. For example, if these dribbling efficiencies are uniformly spread out in the [0, 1] range, then it is likely that a more accurate estimate of the opponent's efficiency can be obtained. Also, if the opponent's actual efficiency is the same as that of one of the teammates of the learning agents, then the estimate will be very close to the actual value. On the other hand, if the efficiencies of the teammates of the learners were all similar, e.g., all efficiencies in a small range [0.6, 0.8], then the learning agents' estimate is likely to be erroneous if the opponent's actual efficiency is quite different. The purpose of estimating the opponent's efficiency is to find out a more effective way of countering it. Fortunately, we can develop adaptation strategies which can tolerate some error in the estimation process. We are currently working on developing such robust group adaptation mechanisms.

3.3 Estimating proper shooting distance

In this section we present a learning scenario where an agent is trying to estimate an upper bound of the distance from the opponent goal from which it can consistently score by shooting. The agent uses its on-line experience to calculate this distance.

At the outset, the agent tries to shoot to score a goal any time it has possession of the ball within a distance δ (also called shooting distance) from the opponent goal. The agent keeps a record of each shooting instance. The record consists of shooting distance and result of shooting action. The agent is initially given a shooting distance. Over repeated shooting experiences it changes δ to more accurately reflect its actual capability. The player updates its shooting distance depending on its success or failure.

If a player is successful in scoring a goal from a distance d, then it considers increasing δ by $\beta * (\delta - d) + \alpha\delta'$, where α and β are learning and momentum constants and δ' is the last change made to δ. Before it actually increments the shooting distance, though, it retrieves the success percentage of shooting (percentage of time a shot taken from a distance less than or equal to this from the goal resulted in a goal being scored) from this new distance. This data may have been gathered at an earlier time in the experiment (the estimate will increase and decrease several times before it settles with very little fluctuations). If no previous information for that shooting distance is available, the player increases δ as above. If prior shooting success information is available, then δ is increased only if the loss of observed success percentages from the current position and the new position is less than $k\%$ (we have used $k = 10$). This means that the player is not willing to increase the shooting distance when there is a significant loss in the success percentage when shooting from the new distance.

An analogous update is performed to reduce δ when a shot by the player is blocked by the opponent goalkeeper. In this case, the shooting distance is not reduced if the reduction does not result in a significant increase of the success percentage of scoring a goal.

The results from experiments with this learning mechanism are very encouraging. Figure 5 shows the estimation of distance when shooting efficiency is constant and goalkeeping efficiency is varying and vice versa. The agents percentage of success and percentage of shooting decision for varying shooting efficiency when goalkeeping efficiency is kept constant are tabulated in Table 3. Same values when goalkeeping efficiency is varying and shooting efficiency is kept constant are depicted in Table 2. The figures and tables clearly shows that the learner converges to a shooting distance from which it has considerable success in scoring goals. It also decides to shoot in very selective situations as is to be expected. The trend is observed in a variety of settings of goalkeeping and shooting efficiencies. As such, the learning mechanism appears to be quite robust.

4 Continuing work

Our implemented system have been tested with relatively simple individual and group behavioral strategies. Currently we have all the players following the same strategies. In the next step we will have only the members of the same team

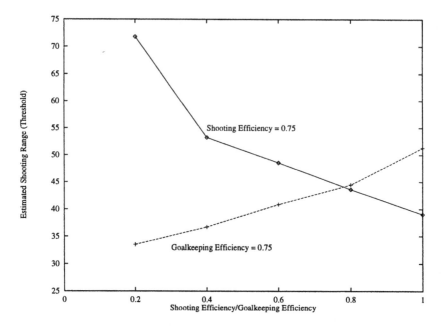

Fig. 5. Variation of estimated distance from which a forward will attempt scoring. For constant Shooting Efficiency the shooting distance decreases as Goalkeeping Efficiency increases. For constant Goalkeeping Efficiency the shooting distance increases as Shooting Efficiency increases.

GoalKeeping Efficiency	percentage of decision to shoot	percentage of success
0.2	11.56	84.37
0.4	9.14	79.65
0.6	8.41	77.13
0.8	7.84	74.75
1.0	7.45	71.84

Table 2. Table of successful shoot percentages and shooting decision percentages for varying Goalkeeping Efficiency and constant Shooting Efficiency.

sharing strategies, and in the long run, each player may be given individual strategies just as it is given individual skill levels. At this stage only the team strategies need to be initially shared among teammates. We currently run the system for a fixed number of time ticks of the simulator.

Currently, agents learn only during the course of a game, but we plan to include long term memories not only about the agents' own skills and how they correlate with performance (e.g., passing efficiency with passing accuracy), but also about the models of other players it has played with in the past. The lat-

Shooting Efficiency	percentage of decision to shoot	percentage of success
0.2	6.48	67.31
0.4	7.18	70.51
0.6	7.54	72.71
0.8	8.11	75.85
1.0	8.91	78.65

Table 3. Table of successful shoot percentages and shooting decision percentages for varying Shooting Efficiency and constant Goalkeeping Efficiency.

ter assumes that every agent introduced into the system should be uniquely identifiable. Our current decision mechanism for choosing to pass or not is very conservative. We plan to investigate a parameterized risk taking scheme that will prompt an agent to pass if the probability of a successful pass is above a given threshold. This threshold can actually varied by agents during the course of the game. For example, agents may decide to take more risky actions when they are losing and time is running out.

The major research issues we want to investigate is how to adapt the team strategies to improve the performance. This include re-assignment of roles, forming better models of opponents by sharing information, pro-actively creating circumstances that will generate necessary information for learning that would otherwise not be available otherwise (e.g., pass the ball to the right forward to test the skill level of the opponent left back), learning to take calculated risks (e.g., rather than passing the ball back to an unguarded teammate, pass it forward to a better guarded teammate who is closer to the opponent goal and thus has a better chance of scoring a goal), etc.

We also plan to investigate a layered group architecture where the entire team is divided into sub-groups (like forwards, defenders, etc.), and subgroups are further divided into individual agents. We will evaluate the effectiveness of different commitment schemes that allow agents to predict the coarse-level behavior of teammates but will still allow individual agents to maintain considerable local flexibility to fulfill its social commitments.

Acknowledgments This research has been sponsored, in part, by the National Science Foundation under a Research Initiation Award IRI-9410180.

References

1. M. Asada, E. Uchibe, S. Noda, S. Tawaratsumida, and K. Hosoda. Coordination of multiple behaviors acquired by vision-based reinforcement learning. In *Proc. of International Conference on Intelligent Robots and Systems*, pages 917–924, 1994.
2. Jerry Banks and II John S. Carson. *Discrete-Event System Simulation*. Prentice-Hall, 1984.

3. Anne Collinot, Alexis Drogoul, and Philippe Benhamou. Agent oriented design of a soccer robot team. In *Proceedings of Second Internationa Conference on Multi-agents Systems (ICMAS'96)*, pages 41–47, Menlo Park, CA, 1996. AAAI Press.

4. David Flanagan. *Java in Nutshell*. O'Reilly, 1996.

5. Steve Hanks, Martha E. Pollack, and Paul R. Cohen. Benchmarks, test beds, controlled experimentation, and the design of agent architectures. *AI Magazine*, 14(4):17–42, Winter 1993.

6. William W. Hines and Douglas C. Montgomery. *Probability and statistics in engneering and management science, 2nd ed.* John Wiley & Sons, New York, NY, 1980.

7. L.P. Kaelbling, Michael L. Littman, and Andrew W. Moore. Reinforcement learning: A survey. *Journal of AI Research*, 4:237–285, 1996.

8. Hiroaki Kitano, Minoru Asad, Yasuo Kuniyoshi, Itsuki Noda, and Eiichi Osawa. Robocup: The robot world cup initiative. In *Working Notes of the IJCAI-95 Workshop on Adaptation and Learning in Multiagent Systems*, August 1995.

9. Hitoshi Matsubara, Itsuki Noda, and Kazuo Hiraki. Learning of cooperative actions in multiagent systems: A case study of pass play in soccer. In Sandip Sen, editor, *Working Notes for the AAAI Symposium on Adaptation, Co-evolution and Learning in Multiagent Systems*, pages 63–67, Stanford University, CA, March 1996.

10. Robocup: The robot world cup. URL: http://www.csl.sony.co.jp/person/kitano/RoboCup/RoboCup.html. Announcement for the first RoboCup competition to be held at IJCAI-97.

11. D.E. Rumelhart, G.E. Hinton, and R.J. Williams. Learning internal representations by error propagation. In D.E. Rumelhart and J.L. McClelland, editors, *Parallel Distributed Processing*, volume 1. MIT Press, Cambridge, MA, 1986.

12. Peter Stone and Manuela Veloso. Collaborative and adversarial learning: A case study in robotic soccer. In Sandip Sen, editor, *Working Notes for the AAAI Symposium on Adaptation, Co-evolution and Learning in Multiagent Systems*, pages 88–92, Stanford University, CA, March 1996.

13. Philip D. Jr. Straffin. *Topics in the theory of voting*. The UMAP expository monograph series. Birkhauser, Boston, MA, 1980.

Learning Agents' Reliability Through Bayesian Conditioning: A Simulation Experiment

Aldo Franco Dragoni, Paolo Giorgini

Istituto di Informatica, Università di Ancona,
via Brecce Bianche, 60131, Ancona (Italy)
{dragon,giorgini}@inform.unian.it

Abstract. This paper reports the first results of a simulation experiment. There are two databases, one containing true propositions and the other containing their respective negations. Five agents in turn access one of them. Each agent has a "capacity" that will be used as the frequency with which the agent accesses (unconsciously) the database with the correct knowledge. Agents randomly exchange information with the others. Since they have limited degrees of capacity, their "cognitive state" quickly becomes inconsistent. Each agent is equipped with the same belief revision mechanism to detect and solve these contradictions. This adopts the Dempster's Rule of Combination to evaluate the credibility of the various pieces of information and Bayesian Conditioning to estimate the relative degrees of reliability of the agents (itself included). The purpose of the experiments was that of evaluating on a statistical basis, the emergent cognitive behavior of the group.

1 Introduction

A biologist, asked whether there will ever be on earth organisms more complex than human brain, answered that such an astonishing super-brain already exists and it is the community of the minds interacting over the surface of the globe. Extending this idea, one might argue that these interacting minds are not necessarily embodied in currently living brains, since amanuenses and printing made it possible the interaction (monodirectional) among intelligence incarnated in brains distant many generations. Present technology greatly improved our ability to interact over time and space; by multiplying and enforcing the links among individuals, it contributed to the development of the global mind mentioned by the biologist. In our days, technology overwhelmed itself by creating artificial (non-biological) bodies in which it is possible to materialize intelligence (intended as the ability to conceive and solve a problem) and communicative abilities. So, the global mind has to be conceptually extended to comprise also such artificial brains.

We concentrate on a very specific problem of collective intelligence: the *distributed elicitation* of the knowledge, i.e., how is it possible that from an enormous variety of inferential schemes and judging capabilities, from different opinions and dogmas, from distinct perspectives and opposite point of views, after a continual interaction, a more uniform (if not unique) vision of the world emerges? Under which singular circumstances the emergent representation of the world results a correct one? The idea is to capture some successful mechanisms in order to replicate them in a world of intelligent highly-engineered (programmed or trained) interacting agents.

In this regard, a central question is how should each agent ascribe a relative degree of reliability to any agent (itself included). Indeed, one of the main differences between classical and multi-agent learning is that, while it makes no sense saying that the external world one should learn about is *incorrect*, the agents from which one get information can do lye, deliberately or not. Even the sensing equipment from which one experiences the external world may be faulty, thus, even for a collaborative group, it is important to define methods to assess each member's relative degree of reliability. According to us, these ascription should be performed under "liberal policies", i.e., each one should be permitted to stand on its own opinion regarding him/herself and the others. Furthermore, these ascription should be performed on the basis of the reciprocal experience and acquaintance, thus we could talk of a form of learning *about* the others in addition to learning *from* the others.

With the distributed elicitation of the knowledge we distinguish two desiderata:

- *convergence*: which local decision strategies favor the convergence of the opinions rather than the chaotic divergence or the static indifference?
- *correctness*: among the local decision strategies of the previous point, which favor the convergence to the most correct pieces of knowledge globally available to the overall agency?

By "local strategies" we mean:

- policies of communication
- methods to form a personal opinion

Each agent's policy of communication should define:

C1. *when* communicate

C2. *what* communicate

C3. to *whom* communicate

These criteria are almost conscious to humans. On the contrary, the ways we form our opinion from:

- information directly perceived from our experience of the world
- information received from the others

seem quite unconscious. Perhaps, one makes at least three kinds of check after the incoming of a previously unbelieved information:

A1. "although I was not aware of it, is this information in accordance with my direct experience of the world?"

A2. "how many people I know believe in it?"

A3. "how much reliable are those people?"

People who prefer the check A1 are confident in themselves; those who prefer A2 are rather conformist, while check A3 is preferred by the suspicious people.

A question studied in this paper is: if all the individuals adopt the same local criteria to evaluate an incoming information, how the global process of knowledge elicitation is affected? For instance, if all the individuals adopt only the control A1, then we would expect that no gain results from the interaction and each remains in his personal degree of correctness. On the other hand, if all the individuals adopt exclusively A2, with no regard toward his own and the others' reliability, then we would expect a global flattening to the medium degree of correctness of the agency. Perhaps we should adopt a reasonable mix of these criteria.

Unfortunately, these criteria are very rough to be studied on a statistical simulation basis. We need more precise techniques. Vaguely, opinion formation deals with:

O1. the way we integrate evidences coming from different sources (ourselves included)

O2. the way we estimate the reliability of the various sources (ourselves included)

O3. the way we revise our opinions after the incoming of a new information that contradicts some of our previous convictions

In section 2 we introduce the techniques our agents adopt to perform O1÷O3. Section 3 introduces the simulation experiment that we are currently carrying on, and reports some of the first results. We will conclude presenting our future work in section 4.

2 Revising/Integrating Knowledge in a Multi-Source Environment

As computer scientists, we are more interested in computer networks than in philosophy, sociology or history of science, so, from now on, we leave the generic distributed scenario for that of a net of artificial (programmed), intelligent and interacting agents.

As the importance of computer networking increases, some tasks traditionally performed by a single workstation are being conceived as performable in a distributed manner by groups of interacting knowledge-based systems. Sometimes, the contribution coming from the other systems could be harmful rather than helpful, for instance because:

B1. local knowledge available at the other systems may not be updated or truthful

B2. there can be some hardware faults at the other workstations

B3. there can be some mistakes in the architecture or software implementation of the other systems

B4. some nodes could join the network with non-cooperative intentions and could supply incorrect information in order to induce wrong decisions and inferences

B5. even if the systems adopt standard protocol and syntax to exchange information, there can still be some semantic differences or ambiguities

B6. there can be noises on the communication channels.

We define as *incompetent* a system that is unreliable because of B1, B2 and B3, and we define as *insincere* a system that is unreliable because of B4. Items B5 and B6 regard communication. We expect that standardization and technology will reduce the effects of B5 and B6; they yield incompleteness rather than incorrectness, i.e., they simply make it impossible to transfer some pieces of information but very seldom cause the reception of a valid piece of information different from the one sent. We concentrate on the kinds of mistakes B1÷B4, which can be ascribed to the responsibility of the sending software/hardware agent.

Many pieces of incorrect knowledge run silently through the network, spreading from node to node. However, occasionally, a symptom that some degree of incompetence or insincerity is insinuated into the network is the appearance of contradictions in the knowledge base of some nodes. Thus, we need a method to integrate information coming from various sources with limited degrees of correctness. From our experience of multi-agent [1] and investigative domains [2], we choose the Dempster's Rule of Combination to perform O1 (previous section), the Bayesian Conditioning to perform O2 and the "Assumption-Based approach" to perform O3. Starting from the beginning, let us recapitulate here our ideas about how to revise/integrate beliefs in a multi-agent environment.

Initially defined as a symbolic model-theoretical problem [3-6], *belief revision* has also been approached both as a qualitative syntactic process [7,8] and as a numerical mathematical issue [9]. Beliefs can be represented either as weighted sentences of a decidable language L or as sets of weighted possible worlds (the models of the sentences). Weights can be either *reals* (normally between 0 and 1), representing explicitly the credibility of the sentences/models, or *ordinals*, representing implicitly the believability of the sentences/models w.r.t. the other ones. Essentially, belief revision consists in the redefinition of these weights in the light of the incoming information (see [16] for an overview). Most of the symbolic and numerical models for belief revision developed so far obey the following three rationality principles:

P1. *Consistency*: revision must yield a consistent cognitive state

P2. *Minimal Change*: revision should alter as less as possible the previous opinion

P3. *Priority of Incoming Information*: the revised cognitive state must embody the information that caused revision.

While the last principle is acceptable when updating the representation of an *evolving* world, it is not generally justified when revising the representation of a *static* situation. In this case, the chronological sequence of the informative acts has nothing to do with their credibility or importance. Another point is that changes should not be irrevocable. To make practical and useful belief revision in a multi-agent environment, we substitute the priority of the incoming information with the following principle [10].

P3. *Recoverability*: any previously believed information item must belong to the current cognitive state if it is consistent with it.

We will achieve recoverability by imposing the *maximal consistency* of the revised cognitive state.

Beliefs introduced in the agent's cognitive state directly by the various sources (itself included) are called *assumptions*. Those deductively derived from the assumptions (internally to the agent's cognitive state) are called *consequences*. We call *Knowledge Base* (*KB*) the set of the assumptions, and *Knowledge Space* (*KS*) the set of all the beliefs (assumptions + consequences). *KB* and *KS* grow monotonically since none of their element is ever erased from memory. Normally both contain *contradictions*. Essentially, our method of belief revision consists of five steps:

S1. Detection of all the minimally inconsistent subsets of KB (*nogoods*)
S2. Detection of all the maximally consistent subset of KB (*goods*)
S3. Defining a credibility ordering $\leq_{KB}$ over the assumption in KB
S4. Extending $\leq_{KB}$ into a credibility ordering $\leq_G$ over the goods detected in S2
S5. Selecting the preferred good with its corresponding set of consequences

2.1 S1 and S2

S1 and S2 deal with consistency and work with the symbolic part of the beliefs. Their task is that of providing the maximal consistency of the revised cognitive state; they adopt an ATMS [11] as basic mechanism. Goods and nogoods are detected through the Reiter's set-covering algorithm for model-based diagnoses [18].

2.2 S3

S3 deals with uncertainty and works with the numerical part of the beliefs. Its task is that of integrating evidences coming from different sources; the basic mechanism is the Dempster's Rule of Combination. We refer to [12-14] for a complete presentation of the Dempster-Shafer Theory of Evidence. Here we recapitulate the main concepts and definition as they have been exploited in our agents.

To begin with, we introduce two data structures: the *reliability set* and the *information set*. Let $S=\{S_1,...,S_n\}$ be the set of the sources and $I=\{I_1,...,I_m\}$ be the set of the information items given by these sources. Then:

- *reliability set* = $\{<S_1, R_1>,...,<S_n, R_n>\}$, where R_i (a real in [0,1]) is the reliability of S_i, interpreted as the "a priori" probability that S_i is reliable.
- *information set* = $\{<I_1, Bel_1>,...,<I_m, Bel_m>\}$, where Bel_i (a real in [0,1]) is the credibility of I_i.

The reliability set is one of the two inputs of the belief-function formalism (see figure 1). The other one is the set $\{<S_1, s_1>,...,<S_n, s_n>\}$, where s_i is the subset of I made of all the information items given by S_i. The information set is the main output of the belief-function formalism. Let us see now how the mechanism works. Let Ξ denote the set of the atomic propositions of L. The power set of Ξ, $\Omega=2^\Xi$, is called *frame of discernment*. Each element ω of Ω is a "possible world" or an "interpretation" for L (the one in which all the propositional letters in ω are true and the others are false). Given a set of sentences $s\subseteq I$ (i.e., a conjunction of sentences), $[s]$ denotes the interpretations which are a model for all the sentences in s.

The key assumption with this multi-source version of the belief function framework is that *a reliable source cannot give false information, while an unreliable source can give correct information*; the hypothesis that "S_i is reliable" is compatible only with $[s_i]$, while the hypothesis that "S_i is *unreliable*" is compatible with the entire set Ω. Each S_i gives an *evidence* for Ω and generates the following *basic probability assignment* (*bpa*) m_i over the elements X of 2^Ω:

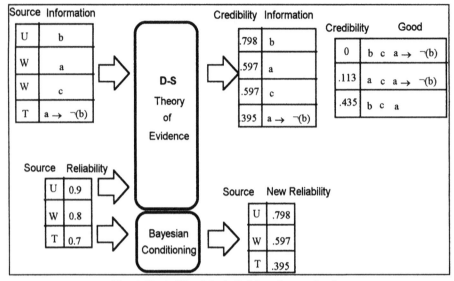

Fig. 1. Basic I/O of the belief-function mechanism

$$m_i(X) = \begin{cases} R_i & \text{if } X = [s_i] \\ 1 - R_i & \text{if } X = \Omega \\ 0 & \text{otherwise} \end{cases}$$

All these *bpa*s will be then combined through the Dempster Rule of Combination:

$$m(X) = m_i(X) \otimes \ldots \otimes m_n(X) = \frac{\sum\limits_{X_1 \cap \ldots \cap X_n = X} m_1(X_1) \cdot \ldots \cdot m_n(X_n)}{\sum\limits_{X_1 \cap \ldots \cap X_n \neq \varnothing} m_1(X_1) \cdot \ldots \cdot m_n(X_n)}$$

From the combined *bpa m*, the credibility of a set of sentences *s* is given by:

$$Bel(s) = \sum\limits_{X \subseteq [s]} m(X)$$

This is the technique adopted to perform O1 in section 1.

Learning each others' reliability

In figure 1, we see another output of the mechanism, obtained through Bayesian Conditioning: the set $\{<S_1, NR_1>, \ldots, <S_n, NR_n>\}$, where NR_i is the new reliability of S_i.

Following Shafer and Srivastava, we defined the "a priori" reliability of a source as the *probability* that the source is reliable. These degrees of probability are "translated" into belief-function values on the given pieces of information. However, we may also want to estimate the sources' "a posteriori" degree of reliability from the cross-examination of their evidences. To be congruent with the "a priori" reliability, also the "a posteriori" reliability must be a *probability* value, not a belief-function one. This is the reason why we adopt the Bayesian Conditioning instead of the Theory of Evidence to calculate it. Let us see in detail how it works here.

Let us consider the hypothesis that only the sources belonging to $\Phi \subseteq S$ are reliable. If the sources are independent, then the probability of this hypothesis is $R(\Phi) = \prod_{S_i \in \Phi} R_i \cdot \prod_{S_i \notin \Phi} (1 - R_i)$. We could calculate this "combined reliability" for any subset of S. It holds that $\sum_{\Phi \in 2^S} R(\Phi) = 1$. Possibly, the sources belonging to a certain Φ cannot all be considered reliable because they gave contradictory information, i.e., a set of information items s such that $[s] = \varnothing$. In this case, the combined reliabilities of the remaining subsets of S are subjected to the Bayesian Conditioning so that they sum up again to "1"; i.e., we divide each of them by "1- $R(\Phi)$". In the case where there are more subsets of S, say $\Phi_1, \ldots, \Phi_l$, containing sources which cannot all be considered reliable, then $R(\Phi) = R(\Phi_1) + \ldots + R(\Phi_l)$. We define the revised reliability NR_i of a source S_i as the sum of the conditioned combined reliabilities of the "surviving" subsets of S containing S_i. An important feature of this way to recalculate the sources' reliability is that if S_i is involved in contradictions, then $NR_i \leq R_i$, otherwise $NR_i = R_i$.

This is the technique adopted to perform O2 in section 1.

2.3 S4

S4 also deals with uncertainty, but at the goods' level, i.e. it extends the ordering $\leq_{KB}$, defined at S3 on the assumptions, into an ordering $\leq_G$ onto the goods.

The belief-function formalism is able to attach directly a degree of credibility to any good g, bypassing S4 in our framework (see again Fig. 1). The problem is that if a good contains only part of the information items supplied by a source, then its credibility is null (see [1] for an explanation). Unfortunately, the event is all but infrequent, so that often the credibility of all the goods is null. Thus, to perform S4 we tried the "best-out" method [8] and the "average" method.

- *best-out method.* Let G' and G" be two goods and let g' and g" be the most credible assumptions (according to $\leq_{KB}$), respectively, in KB\G' and KB\G". Then $G'' \leq_G G'$ iff $g' \leq_{KB} g''$.
- *average method.* The goods are ordered according to the average credibility of their elements

A main difference between the two is that using the *average credibility* method the preferred good(s) may no longer necessarily contain the most credible belief(s).

2.4 S5

The good with the highest priority in $\leq_G$ will be selected in S5 as the preferred good. In case of ties, the revised cognitive state may be based on either one of the goods with the same highest priority (randomly selected) or on their intersection (see [8] and [15]). This latter case means rejecting all the conflicting but equally credible information items. The result is not a good (it is not maximally consistent) and thus implies rejecting more assumptions than necessary to restore consistency.

3 The Simulation Experiment

The task given to the group is that extracting as much correct knowledge as possible from a corrupted knowledge repository. The knowledge repository is trivially implemented as a couple of databases containing the same quantity of information items: a database holds correct information, the other one contains the negations of the information items in the correct database. Agents cannot distinguish the two databases. Each agent is characterized by a degree of "capacity" (between 0 and 1) that will be adopted as the frequency with which it accesses (unconsciously) the correct database.

While learning from each others, agents exchange information, thus, they should be equipped with a communication module (Comm) specifying the three basic communication policies C1÷C3 (section 1). These policies will affect deeply the global cognitive behavior of the group. We adopted the following rules:

H1. the communication is peer-to-peer and the recipient of the communication is selected randomly (C3)

H2. the piece of information to communicate is chosen randomly among those in the preferred good (C2)

H3. agents communicate (and access the databases) once per simulation cycle (C1)

H4. agents do not communicate to the others the sources from which they received the pieces of information, but they present themselves as completely responsible for the knowledge they are passing onto the others; the receiving agents consider the sending agents as the sources of the information they are receiving

H5. agents exchange opinions regarding the credibility of the information they are providing to the others, but they do not exchange opinions regarding the reliability of the other agents with whom they got in touch

With H4 we extend the scope of responsibility: an agent is responsible not only for the information that it provides to the network as the original source, but also for the information that it receives from other agents and, *retaining it credible*, passes it on to the others. With H5 we limit the range of useful information: an agent's opinion regarding the others' (and its own) reliability is drawn out from pure information regarding the knowledge domain under consideration, not from indirect opinions.

For each agent we evaluated three parameters: its average *reliability*, the *quality* and the *quantity* of the beliefs in its preferred good.

Quality and quantity are evaluated as differences w.r.t. the case without interaction.

- $Quality = Q - Q_{\text{without communication}}$

  where $Q = \dfrac{\left|\text{true propositions in G}\right| + \left|\text{false propositions outside G}\right|}{\left|\text{propositions in KB}\right|}$

- $Quantity = \left|\text{true propositions in G}\right| - \left|\text{true propositions in G}\right|_{\text{without communication}}$

- *Reliability* of the ith agent $R_i = \dfrac{\sum\limits_{k=1}^{|agents|} r_{ki}}{|agents|}$ where r_{ki} is the reliability of the ith

agent, estimated by the kth one.

Due to severe complexity and duration problems, we fixed at five the number of agents. Each of the two databases contains 100 propositions. From some tentative with 40 accesses for each agent, we realized that 25 accesses were sufficient to reach stable values. To reduce the effects of casuality, each of the sixteen simulations has been repeated 20 times. We calculated the average of each parameter, for each cycle, over the twenty repetitions.

3.1 Experiment 0

In order to evaluate the performances of Bayesian Conditioning to estimate the reliabilities, we made a series of simulations focusing our attention on the perspective of a single agent. This agent receives information from four agents with different degree of capacity. If some of them have a degree of capacity clearly less than the others, Bayesian Conditioning is able to detect the less capable agents (see fig 2; case with only one agent with a very low degree of capacity).

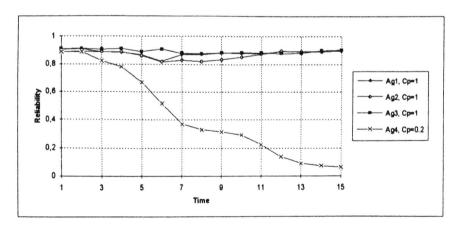

Fig. 2

More generally, when the agents' capacity are sufficiently differentiated (with differences greater than 0.2) there is an acceptable correspondence between estimated reliability and real capacity (see the case in fig 3), but this correspondence is lost when the degrees of the agents' capacity are closed to each others (see the case in fig 4).

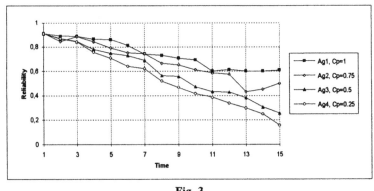

Fig. 3

Fig. 4

Experiment 0 showed that, for our purposes, Bayesian Conditioning was an acceptable way to estimate the degrees of reliability of information sources affected by limited degrees of correctness (capacity, competence, sincerity ...).

3.2 Experiment A

We began the "distributed" experiments by making two series of simulations. In the first one we gave to the five agents the same capacity. We made six simulations, from capacity 0 to capacity 1 (fig. 5). In the second series (ten simulations) we gave to the agents the capacities reported in fig. 6.

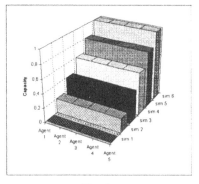

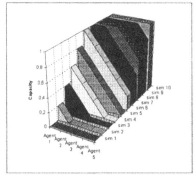

Fig. 5   **Fig. 6**

The results obtained can be summarized as follows.

Quality

Interaction increases the quality of an incapable agent's cognitive state and decreases the quality of a capable one. A typical trend is reported in fig. 7, which refers to the sim 5 in fig. 6 (decreasing capacity from 1 to 5). The average quality (Av) stands at zero. This means that if there are few incapable agents, then they gain much in correctness while the others lose very little.

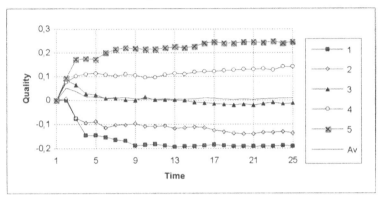

Fig. 7

Quantity

Interaction always increases quantity. However, incapable agents gain more than the capable ones. A typical trend is reported in fig. 8 (sim 5 fig. 6; decreasing capacity from 1 to 5).

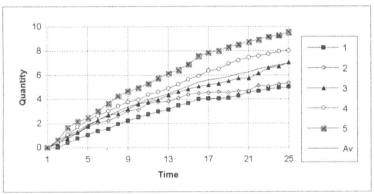

Fig. 8

Reliability

Reliabilities follow typical trends as the one reported in fig. 9 (sim 7 in fig. 6; decreasing capacity from 1 to 5). We see that all the agents lose reliability, but there is what we called "majority effect". If the average capacity of the group is greater than 0.5, then the capable agents lose less than the incapable ones.

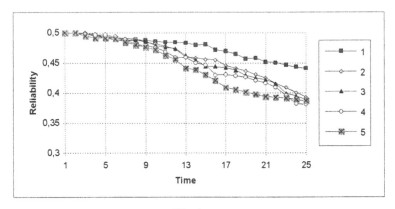

Fig. 9

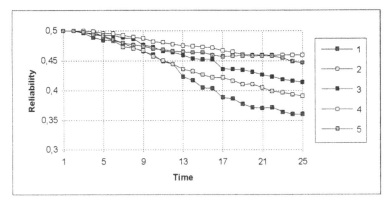

Fig. 10

On the opposite, if the average capacity is less than 0.5, than the capable agents lose more than the incapable ones, as in fig. 10 (sim 3 fig. 6). In this case, since the capable agents are in minority, they are regarded as unreliable by the others. The result is that the estimated reliability is quite inversely proportional to the real capacity.

These results were expected, since this group looks like a class without teacher, or a scientific community without experimental evidence. They are positive in the sense that the members of the group acquire knowledge without loosing (averagely) in correctness.

3.3 Experiment B

Subsequently, we introduced a "teacher" agent with the following features:

- capacity = 1 (it accesses only the correct database)
- for each k, $r_k = 1$ (all the agents know that the teacher is absolutely reliable)
- it transmits but doesn't receive information (the teacher's cognitive state will not be contaminated by the others)

We repeated the simulations with the teacher and five agents as in the experiment A. Results may be summarized as follows.

Quality

The main difference w.r.t. the case without teacher is that now the average quality is positive and lightly increasing. This is due to the presence of the teacher which, occasionally (communications took place randomly) gave to the others the possibility to choose in the right way. A typical trend is showed in fig. 11 (capacity decreases from agent 1 to agent 5).

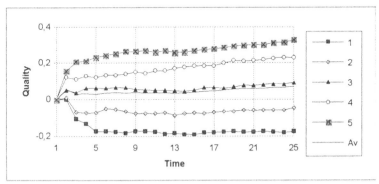

Fig. 11

Quantity

Even in this case, interaction increases the amount of correct data in the cognitive state of every agent, and the gain is inversely proportional to the capacity of the agent. The effect of the teacher is that the gain is higher than before for any agent. A typical trend is showed in fig. 12 (decreasing capacity from 1 to 5).

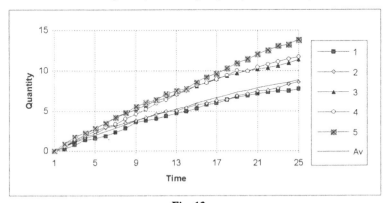

Fig. 12

Reliability

The correspondence between estimated reliability and real capacity holds when the average capacity is greater than 0.5. A typical trend is showed in fig. 13 (decreasing capacity from 1 to 5). Under that value the situation becomes confused. What we called "majority effect" in the experiment A is no longer appreciable when there is a teacher in the group.

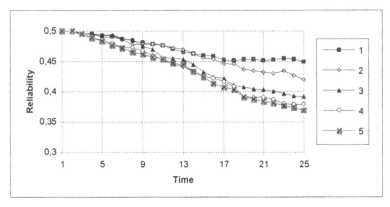

Fig. 13

3.4 Experiment C

In the previous experiments, the characteristics of the agents were static; their capacity did not change during the simulation. We were curious to see if the group would have been able to realize that some of its member changed their degree of capacity, and how long would have it taken to the group to be aware of the change. After several simulations with a agent decreasing or increasing its capacity, we realized that only high quality groups were able to perceive the change. For groups with an average capacity less than 0.6, the situation at the time of the change was sufficiently chaotic to hide it. This implies that only decreases of capacity will be perceived by the group. The graph in fig. 14 reports the "quality" trends of five agents with capacity 0.98, where the fifth one, at the 15th simulation cycle, decreased its capacity to 0.8.

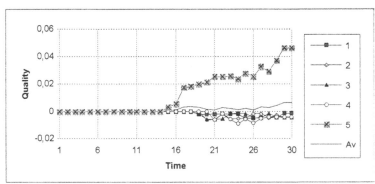

Fig. 14

The graph in fig. 15 reports the "quantity" trends of five agents with capacity 0.8 where the fifth one, at the 15th simulation cycle, decreased its capacity to 0.6.

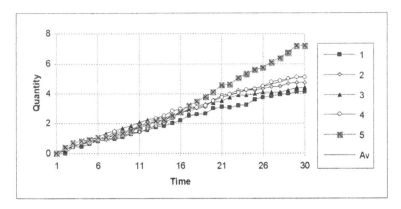

Fig. 15

The graph in fig. 16 reports the "reliability" trends of five agents with capacity 0.98 where the fifth one, at the 15th simulation cycle, decreased its capacity to 0.8.

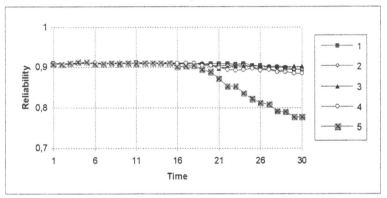

Fig. 16

As an example of a low quality group, fig. 17 reports the "reliability" trends of five agents with capacity 0.6 where the fifth one, at the 15th simulation cycle, decreased its capacity to 0.4. It is evident that the group was not able to perceive the change.

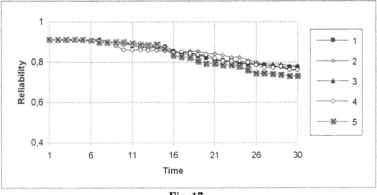

Fig. 17

We also tried to study the effects of the deepness of the reciprocal acquaintance: the longer the period spent together, the more reluctant should be the agents in changing their opinions regarding their companion who decreased its capacity. As a matter of fact, fig. 18 shows the "reliability" trend of the agent who changed its capacity from 0.98 to 0.8 at different time points: t=0, 5, 10, 15, 20 and 25. We can see that the later the change, the lighter the curve's slope: the group which appreciated the agent's reliability for a longer period has more inertia to change its opinion regarding that agent.

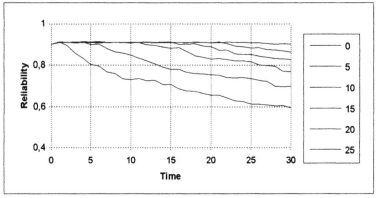

Fig. 18

4 Conclusions

Learning other agents' (and its own) relative degrees of reliability is crucial when interacting within a group of partially competent (correct, sincere, etc.) information sources. The technique proposed in this paper is a simple application of Bayesian Conditioning. Results have been extracted from the first outcomes of a broader minded simulation experiment: studying the global cognitive behavior of a group whose members adopt definite local mechanisms to revise their opinions (regarding agents and knowledge), local policies of interaction and global strategies of integration. The local method to revise beliefs is a form of assumption-based reasoning which adopts the Dempster's Rule of Combination to integrate, information coming from different sources, and embodies Bayesian Conditioning to evaluate the reliability of those sources. Our feeling is that this latter technique is very intuitive and simple. However, while it shows interesting properties from a distributed perspective (the global cognitive behavior of the group seems sociologically intuitive and foreseeable), we still need to deepen into its performances as a centralized mechanism to evaluate the reliability of partially correct information sources (e.g., sensors). We will soon report in another paper the results of a specific simulation experiment that we are still carrying on.

References

[1] A.F. Dragoni, P. Giorgini, "Belief Revision through the Belief Function Formalism in a Multi-Agent Environment", Third International Workshop on Agent Theories, Architectures and Languages, LNAI Series, Springer-Verlag, 1997.

[2]   Dragoni A.F., Ceresi, C. and Pasquali, V., A System to Support Complex Inquiries, in Proc. of the "V Congreso Iberoamericano de Derecho e Informatica", La Habana, 6-11 march 1996.

[3]   Alchourrón C.E., Gärdenfors P., and Makinson D., On the Logic of Theory Change: Partial meet Contraction and Revision Functions, in The Journal of Simbolic Logic, 50, pp. 510-530, 1985.

[4]   P. Gärdenfors, Knowledge in Flux: Modeling the Dynamics of Epistemic States, Cambridge, Mass., MIT Press, 1988.

[5]   P. Gärdenfors, Belief Revision, Cambridge University Press, 1992.

[6]   Williams M.A., Iterated Theory Base Change: A Computational Model, in Proc. of the 14th Inter. Joint Conf. on Artificial Intelligence, pp. 1541-1547, 1995.

[7]   W. Nebel, Base Revision Operations and Schemes: Semantics, Representation, and Complexity, in Cohn A.G. (eds.), Proc. of the 11th European Conference on Artificial Intelligence, John Wiley & Sons, 1994.

[8]   Benferhat S., Cayrol C., Dubois D., Lang J. and Prade H., Inconsistency Management and Prioritized Syntax-Based Entailment, in Proc. of the 13th Inter. Joint Conf. on Artificial Intelligence, pp. 640-645, 1993.

[9]   Dubois D. and Prade H., A Survey of Belief Revision and Update Rules in Various Uncertainty Models, in International Journal of Intelligent Systems, 9, pp. 61-100, 1994.

[10] Dragoni A.F., Mascaretti F. and Puliti P., A Generalized Approach to Consistency-Based Belief Revision, in Gori, M. and Soda, G. (Eds.), Topics in Artificial Intelligence, LNAI 992, Springer Verlag, 1995.

[11] de Kleer J., An Assumption Based Truth Maintenance System, in Artificial Intelligence, 28, pp. 127-162, 1986.

[12] Shafer G. and Srivastava R., The Bayesian and Belief-Function Formalisms a General Perpsective for Auditing, in G. Shafer and J. Pearl (eds.), Readings in Uncertain Reasoning, Morgan Kaufmann Publishers, 1990.

[13] Shafer G. (1976), A Mathematical Theory of Evidence, Princeton University Press, Princeton, New Jersey.

[14] Shafer G., Belief Functions, in G. Shafer and J. Pearl (eds.), Readings in Uncertain Reasoning, Morgan Kaufmann Publishers, 1990.

[15] Benferhat S., Dubois D. and Prade H., How to infer from inconsistent beliefs without revising?, in Proc. of the 14th Inter. Joint Conf. on Artificial Intelligence, pp. 1449-1455, 1995.

[16] Dragoni A.F., Belief Revision: from theory to practice, to appear on "The Knowledge Engineering Review", Cambridge University Press, 1997.

[17] A.F. Dragoni, P. Giorgini and P. Puliti, Distributed Belief Revision vs. Distributed Truth Maintenance, in Proc. 6th IEEE Conf. on Tools with A.I., IEEE Computer Press, 1994.

[18] R. Reiter, A Theory of Diagnosis from First Principles, in Artificial Intelligence, 53, 1987.

A Study of Organizational Learning in Multiagents Systems

Masahiro Terabe[1], Takashi Washio[2], Osamu Katai[3], Tetsuo Sawaragi[3]

[1] Mitsubishi Research Institute, Inc.
terabe@mri.co.jp

[2] Institute of Scientific and Industrial Research, Osaka University
washio@isir1.sanken.osaka-u.ac.jp

[3] Graduate School of Engineering, Kyoto University
{katai,sawaragi}@prec.kyoto-u.ac.jp

Abstracts. In this paper, we are concerned with "organizational learning" in the multiagents systems. As an example for the organizational problem solving process, we will take the task allocation process. The process always enhances the performance of organization, however it is difficult for designers to make the process suitable for the organization and its environment. For that reason, the learning ability is necessary for the process, since it gives them adaptability and robustness. This paper is intend to investigate the relation between selection of task allocation style and its task allocation costs in the learning organization. We introduce an organizational learning model consisting of reinforcement learning agents. These agents learn about ability of other agents in the organization and themselves through their experience of interaction. Thus, we show the results of simulation, and discuss on them.

1 Introduction

This paper focuses on the "organizational learning" in the multiagents systems. In the real world, there are many kinds of multiagent systems, such as human organization, economical market system, networked computers, and so on. Many researchers in social science and computer science have started to research on organizational learning since about ten years ago [OL 96] [Gasser 89] [Terano 94]. For the system designer, organizational learning process is interesting because it may give the multiagents systems adaptability and robustness to themselves and their environment. Furthermore, it is meaningful for the human organization managers to understand the mechanism of the organizational learning as human organization is composed by learning agents.

As an example for the part of organizational problem solving process, we will take the task allocation process. The process always enhances the performance of organization, however it is difficult to make the process suitable for the organization and its environment. For that reason, the learning ability is necessary for the process, since it gives them adaptability and robustness. This paper is intend to investigate the relation between selection of task allocation style and task allocation costs in the learning organization.

To analyze the feature, we propose organizational learning model consisting of a kind of reinforcement learning agents. These agents learn about ability of other agents in the organization and themselves through their experience of interaction. We discuss the feature of the process through computational experiments.

2 Research Background

A considerable number of studies have been made on organizational learning. In this section, we will begin with following the related works mainly in social science and computer science (distributed artificial intelligence) as our research background. Additionally, we make our interests clear.

2.1 Related Work in Social Science

Many social science researchers have been interested in organizational learning in the last decade [OS 91]. However the definition of organizational learning has not been cleared. Therefore, Cohen et al. showed a rough map of the territory of organizational learning researches and clarify it as some research dimensions: they are building knowledge types in the learning, changes in organizational dynamics, and the stored place of learning results [OL 96]. In such research dimensions, our research interest belongs to the dynamic process of organization caused by organizational learning. Dynamic process in the organization includes the changes in decision making of agents and communication patterns, in other words a kind of organizational structure defined by them.

Malone investigated the organizational task processing costs caused in each kind of his defined organizational coordination structures. He showed that there is trade off among the production costs, coordination costs, and vulnerability costs [Malone 87]. He reports the result of his analysis as follows. The hierarchy type of organization requires more production costs than decentralized market type organization On the contrary, the communication costs is smaller. Additionally, he suggests that the best organizational structure, which minimized the total costs, is defined by the situation. However he didn't consider the learning perspective in the organization.

In this paper, every agent has learning mechanism and a kind of costs estimation mechanism. The agents calculate the estimation based on its knowledge and information taken through communication. We compose the organization by these agents, and analyze its behavior.

2.2 Related Work in Distributed Artificial Intelligence

There are many researches on the task processing by multiagents systems [Gasser 89]. These systems are composed by some numbers of autonomous human or machine agents, and they sometimes communicate and cooperate to allocate tasks adequately. Smith proposed Contract Net Protocol (CNP) for task allocation among multiagents as a kind of cooperative protocol modeling contract process in the human society [Smith 80]. As its extended work, Ohko et al. adopted CBR to the Contract Net Protocol [Ohko 96]. Their manager agent reasons for the contractor's ability from past experiences, and the organizational structure is self-organized in the learning process.

Weiss investigated a kind of organizational learning [Weiss 94] [Weiss 95]. In his work agents can process only a part of the task, but the part is different among agents. These agents learn on the better task allocation path by reinforcement learning, and as a result of learning, an effective organizational task processing structure emerges.

Our purpose of this paper is also belongs to investigate the relation between organizational learning process and the emergence of organizational structure, however we intend to discuss mainly on the factor of task allocation costs in the organizational task allocation process as mentioned above.

3 Organizational Learning Model

3.1 Organizational Learning and Individual Learning

Individual Learning

In the case each agents learn as a member of the organization, individual learning is roughly divided into two kinds from its learning objects as follows.

- Learning on the environment

Individual agents learn on the environment through interacting with it. The environment includes the other agents that don't belong to the organization.

- Learning on the agents in the organization

Individual agents learn how to cooperate to achieve an organizational goal. This learning is different from *learning on the environment* because the learning agents can communicate with their learning objective, other agents.

Organizational Learning

As the results of agents' individual learning, agent's knowledge, their part in the organization, and their behavior change to achieve the organizational goal more efficiently. This is a kind of organizational learning. In this process, organizational structure defined by information and task flow changes so that it looks as if organization itself learned.

3.2 Effect of the Organizational Learning

Through learning on the environment and the inside of the organization as either individual or organization, a kind of agents' collective and cooperative behavior emerges. Additionally, agents become to expect the others' action based on his knowledge in some extent So, they get to be able to choose the adequate action to achieve the organizational goal even if they would not communicate. This is one of the important effect of organizational learning in the multiagents systems. Multiagents system has more potentiality for the problem solving than single agent system because of its greater amount of resources for problem solving. However, it needs a lot of resources for communication so that system can not use all of the resources for problem solving. Therefore, learning organization can reduce its communication costs so that organization is able to invest their resource to task processing concentricity.

3.3 Task Allocation Problem

Task allocation process is a popular problem in the DAI [Gasser 89]. This problem involves many features of the real world (for example: human organization [Sikora 96], flexible manufacturing systems [Shaw 89], load balancing in the networked computers [Shaerf 95], and etc.). In this paper, we also take this problem to simulate the behavior of learning multiagents organization.

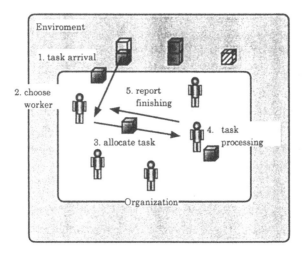

Fig.1. Task allocation problem

The detail of the problem is indicated in Fig. 1. The organizational goal of task allocation is to minimize the total costs for organization consisting of task processing costs and communication costs. Each agent has ability to process a given task. The ability can be used for any kinds of tasks, but the effect for each task is not restricted to be the same. Allocating task to the other agents that can process the task in shorter time makes the total organizational costs smaller, although it needs some communication costs. Each agent has also communication ability to the other agent. They are able to learn on their ability and situation through interaction.

The task allocation process is conducted in the following manner.

[Step 1] Tasks come from the outside of the organization, and arrive at each agent with equal probability.

[Step 2] The agent who receives a new task becomes a *manager* of the task.

[Step 3] The manager agent chooses an agent (*worker*) from all agents (including himself) in the organization. The worker agent should process the received task.

[Step 4] The worker process the task.

[Step 5] When the worker completes the task, he reports his finish to the manager.

For decision making on worker in **Step 3**, we consider next three general styles in the organization.

1. Contract Net Process ("CNP")

The manger agent chooses worker agent through contract net process. In this process, manger and the other agents communicate on (1) task announcement (manager to agents), (2) bidding (agents to manager), (3) awarding (manager to worker), (4) reporting for finish of task processing (worker to manager). Additionally, manager has to wait for the bidding of the other agents.

In this process, organization requires $(2*number\_of\_agent+2)$ units' communication costs and $wait\_bidding\_time$ costs in addition to task processing costs.

2. Based on Knowledge ("Knowledge")

The manger chooses the worker agent based on its knowledge on the other agent. In this process, there are two kinds of communications in the organization. They are (1) awarding (manager to worker) and (2) reporting for finish of task processing (worker to manager). So 2 communication costs are required.

3. Manager process by itself ("By itself")

The manger process the managing task by himself. In this process, no communication is required.

The important point to notice is that the total task processing costs are different among each style : "CNP" needs much communication costs and waiting time for bidding, therefore "By itself" needs nothing on it. On the contrary, "CNP" guarantees that manager found the best worker, while the others don't. Agent in the organization should choose adequate style to reduce the total costs through organizational learning.

3.4 Learning Model

In this paper, each agent has a reinforcement learning mechanism[Watkins 92]. Its purpose is to process the tasks within shorter time. The knowledge of the other member's expected time to process a type of task is represented as follows.

$$et_i(agent_j, task_k)$$

This is a value evaluated by the manager $agent_i$ on the expected time of a worker $agent_j$ to process a kind of task $task_k$.

Furthermore, agents have belief on reliability for their knowledge, it expresses collectives and reliability of agents' knowledge on the members.

$$rel_i(task_k)$$

This value shows the evaluation for reliability for $agent_i$ on $task_k$.

From this kind of knowledge representation, the three types of task allocation process and selecting mechanism of task allocation style are modeled as follows.

Task Allocation Style Selection

First, manager checks their reliability for its knowledge. If the reliability is lower than the threshold, manager selects task allocation style from either "CNP" or "By itself". On the contrary, if the knowledge reliability is high enough, manager selects the style from either "Knowledge" or "By itself". In each case, agent estimate the required costs.

In the case for manager selects "By itself", it estimates the task processing costs based on adding the new managing task to its ongoing tasks under the assumption manager process the task by itself. The total costs are equal to the task processing costs because this style doesn't require communication costs and waiting bid costs.

On the other hands, manager estimate the required costs on each agent under the assumption that manager allocate the task to the member. Manager calculates the other agents' task processing costs based on their knowledge. The estimation of the other agents' total costs calculated as sum of the task processing costs and the extra costs which is required for each task allocation process. The extra costs is calculated as follows.

- extra costs in "CNP" :
$extra\_cost=2(number\_of\_agent+1)*communication\_cost+waiting\_bid\_time$
- extra costs in "Knowledge" :
$extra\_cost=2*communication\_cost$

If the estimated costs for the case manager processed is shorter than one of the other agents, the manager allocates the task to itself. In the other cases, manager selects "CNP" or "Knowledge" as explained above.

Modeling Task Allocation Styles

We introduce models of task allocation styles. They include learning process.

"CNP" Style

We model the each contract net process as follows.

- *Task announcement* : manager announces its new managing task to the all members in the organization through communication, and waits members bid in the regulated time (*waiting_bid_time*).
- *Bidding* : Members received the task announcement estimate the required time for the task processing in the case it undertakes the task. Then members report their bid based on the estimation to the manager.
- *Awarding* : Manager asks the task processing to the best bidder (worker), and reports to the worker.
- *Finishing* : As soon as the worker finishes his task processing, worker reports on it to the manager.

- *Learning* : The knowledge is updated through the following iterative substitution, when the manager asks the worker to solve the task and receives the report of task finishing from the worker.

$$et_i(agent_j, task_k) = (1.0 - \alpha)et_i(agent_j, task_k) + \alpha T$$

T is the time that is consumed by the worker to process the given task. α is a learning rate. If it is near to 1.0, the manager agent changes the knowledge depending on new experiences strongly. Additionally, if the task processing time T is close to the manager's estimated time based on his knowledge, the value of knowledge reliability is increasing.

"Knowledge" Style
Manager selects based on the knowledge, the level of the ability of a worker $agent_j$ for $task_k$ is defined as follows in the manager $agent_i$. We call a parameter n as "exploration rate".

$$pd(agent_j) = \frac{1}{\exp\left\{\dfrac{et_i(agent_j, task_k)}{n}\right\}}$$

If a new $task_k$ arrives at an $agent_i$ they choose a worker from agents in the organization by following the probability given in the next formula.

$$Prob(agent_i) = \frac{pd(agent_j)}{\sum_{agents} pd(a)}$$

If the parameter n is small, the manager's decision making tends to select a worker having the best evaluation in the manager.
The learning process is same as the "CNP".

"By itself" Style
Manager selects itself as worker.

4 Analysis

We analyze the task allocation process in the learning organization consisting of 5 agents prepared in the environment of computer simulation. The time interval of task arriving at the organizations is set to be 20 minutes, and a learning rate α and an exploration rate n are 0.5 and 0.1 respectively.
The time period of a simulation is from 0 minutes to 10000 minutes.

Effect of Task Allocation

First, we confirm the efficiency of task allocation. We set agents' time required to process $task_k$ (k=1,2,3) for $agent_j$ (j=A,B,C,D,E) as depicted in Table 1. As you can see, the ability of each agent on each task is different. Each experiment did five times in the same setting.

Table 1 : The required times of agents for tasks (min.)

Agent	Task 1	Task 2	Task 3
A	5.0	12.0	20.0
B	20.0	5.0	12.0
C	12.0	20.0	5.0
D	5.0	5.0	5.0
E	20.0	20.0	20.0

We set next two cases to confirm the efficiency of task allocation by contract net process as follows.

Case 1 : No task allocation
The manager agent processes the received tasks only by itself.

Case 2 : Task allocation with contract net process
The manager allocates tasks by the contract net process. To make the effected of task allocation types remarkable, the communication costs and waiting time for bidding are ignored in this experiment.

The case 1 and case 2 needed to process the task 6069.6 (min.) and 2500.0 (min.), respectively. From this results, we confirm that allocating task through contract net process is more effective than without task allocation.

Effect of Communication Costs

In the next experiment, we analyze the performance of organization with our modeled learning mechanism. The costs for one communication 0.4 and waiting_bid_time are set to be 0.4 and 4.0 respectively.

Table 2 : The total of required task processing costs

style	Organizational Learning			"CNP" only		
	total cost	proc. cost	comm. cost	total cost	proc. cost	comm. cost
costs	4537.4	3458.8	709.0	7657.1	2500.0	3157.1

As indicated in Table 2. The learning organization reduces its total costs and communication costs, while task processing costs (required time) increases.

Table 3 : The selected task allocation style in the last 50 tasks

communication cost per unit	CNP	Knowledge	By itself
0.2	6.4%	60.0%	33.6%
0.3	6.0%	44.4%	49.6%
0.4	9.2%	39.2%	51.6%

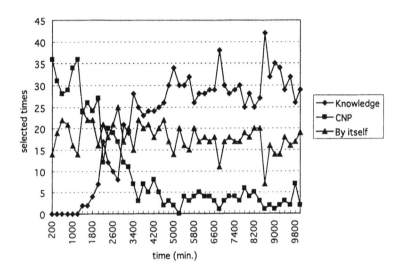

Fig.2. Selected times of each task allocation style

Next we analyze the effect of communication costs to the selection of task allocation style. We simulate three cases and set the one communication costs 0.2, 0.3, and 0.4 respectively. The result is depicted in Table 3. This shows that if the communication costs are high, manager tends to allocate the task to itself more frequently to reduce the communication costs. On the contrary, in the case that the communication costs are low, manager allocates the tasks in "Knowledge" style. At the last stage of the learning, the task flow is self-organized. Therefore, the organized patterns are different. For instance, the allocation of task 1 is concentrated to agent A and agent D if the communication costs are cheap. However, Agent C sometimes process the managing task 1 by itself if the communication costs are high.

Additionally, the selected task allocation style changed through the learning process. At the beginning of the learning, agents allocate most of the tasks by "CNP" style, and after the enough time they change their task allocation style mainly to "Knowledge" style and "By it self" style. This process is drawn in Fig. 2.

5 Discussion

First, from the experiment in the last section, we obtained the result that the total task processing costs are reduced in the learning organization. This fact indicates that the organizational learning increases the efficiency of problem-solving.

The second, the main task allocation style is different in each situation. The manager agents allocate their tasks to themselves more frequently in the case that communication costs are high. On the contrary, manager allocates the other agents based on their knowledge if the communication is cheap.

The third, task and information flow are self-organized through organizational learning. The emerged structures reflect a kind of situation for the organization, communication costs. In this paper, we adopted a simple model of task allocation and processing in an agent organization. However, we get the enough results to know the feature of the organizational learning. The agents with learning mechanism can improve their way of communication and the task allocation.

Finally, as the results of such individual agent's learning, organizational structure defined by the task flow and communication flow changed, and the learned structures reflects the feature of organization and its environment.

Additionally, the following research topics must be worked on in our future work.

- Studies on organizational recognition process and its computational model are required. How individual agents recognize their environment and organization as a member of it? How the organizational members form their common recognition? Organizational recognition process seems to close relation with when and how organization learns.
- Add a knowledge sharing mechanism among agents into the organizational learning model.
- Evaluate the contribution of the diversity of agents' learning strategies to the organizational adaptability and robustness.

6 Conclusions

We introduced an organizational model consisting of reinforcement learning agents on task allocation problem. Through these numerical experiments, we analyzed the relation between communication costs and selection of task allocation style.

References

[OL 96] Cohen, M.D. and Sproull, L.S. eds.: *Organizational Learning*, Sage Publications, California, 1996.

[Gasser 89] Gasser, L. and Huhn, M.N. eds.: *Distributed Artificial Intelligence II*, Morgan Kaufmann, California, 1989.

[Malone 87] Malone, T.W.: Modeling Coordination in Organizations and Markets, *Management Science*, **10**, 1987.

[Ohko 96] Ohko, T., Hiraki, K., and Anzai, Y.: Learning to Reduce Communication Cost on Task Negotiation among Multiple Autonomous Robots, In Sen, S. and Weiss, G. eds., *Adaptation and Learning in Multi-Agent Systems*, pp.177-190, Springer, 1996.

[OS 91] *Organizational Science*, **2** (1), 1991.

[Shaerf 95] Shaerf, A., Shoham, Y., and Tennenholtz, M.: Adaptive Load Balancing: A Study in Multi-Agent Learning, *Journal of Artificial Intelligence Research*, **2**, pp.450-475, 1995.

[Shaw 89] Shaw, M.J., and Whinston, A.B.: Learning and Adaptation in Distributed Artificial Intelligence Systems, In Huhn, M.N. and Gasser, L., eds.: *Distributed Artificial Intelligence II*, pp.119-137, Morgan Kaufmann, California, 1989.

[Sikora 96] Sikora, R., and Shaw, M.J.: A Computational Study of Distributed Rule Learning, *Information Systems Research*, **7** (2), pp.189-197, 1996.

[Smith 80] Smith, R.G.: The contract net protocol: High-level communication and control in a distributed problem solver, *IEEE Transactions on Computers*, **C-29** (12), pp.357-366, 1980.

[Terano 94] Terano, T.: Learning from Problem Solving and Communication : A Computational Model for Distributed Knowledge Systems, *Proc. FGCS'94, Workshop on Heterogeneous Cooperative Knowledge Bases*, p.153, 1994.

[Watkins 92] Watkins, C.J.C.H. and Dyan, P.: Technical Note: Q-Learning, *Machine Learning*, **8**, pp.279-292, 1992.

[Weiss 94] Weiss, G. : Some Studies in Distributed Machine Learning and Organizational Design, Technical report FKI -189-94 TU Muenchen, 1994.

[Weiss 95] Weiss, G.: *Distributed Machine Learning*, Infix, Sankt Augustin, 1995.

Cooperative Case-Based Reasoning

Enric Plaza, Josep Lluís Arcos, and Francisco Martín
IIIA - Artificial Intelligence Research Institute
CSIC - Spanish Council for Scientific Research
Campus UAB, 08193 Bellaterra, Catalonia, Spain.
Vox: +34-3-5809570, Fax: +34-3-5809661
Email: {enric,arcos,martin}@iiia.csic.es
WWW: http://www.iiia.csic.es/Projects/FedLearn/CoopCBR.html

Abstract. We are investigating possible modes of cooperation among homogeneous agents with learning capabilities. In this paper we will be focused on agents that learn and solve problems using Case-based Reasoning (CBR), and we will present two modes of cooperation among them: Distributed Case-based Reasoning (DistCBR) and Collective Case-based Reasoning (ColCBR). We illustrate these modes with an application where different CBR agents able to recommend chromatography techniques for protein purification cooperate. The approach taken is to extend Noos, the representation language being used by the CBR agents. Noos is knowledge modeling framework designed to integrate learning methods and based on the task/method decomposition principle. The extension we present, *Plural* Noos, allows communication and cooperation among agents implemented in Noos by means of three basic constructs: alien references, foreign method evaluation, and mobile methods.

1 Introduction

We are investigating possible modes of cooperation among homogeneous agents with learning capabilities. In this paper we will be focused on agents that learn and solve problems using Case-based Reasoning (CBR), and we will present two modes of cooperation among them: Distributed Case-based Reasoning (DistCBR) and Collective Case-based Reasoning (ColCBR). Before presenting our approach it is relevant to state how we view the relation between cooperation processes and learning processes in the framework of multiagent systems (MAS).

1.1 On Cooperation and Learning

In a multiagent environment, where agents have learning capabilities, the distinction between learning and cooperation is sometimes blurred. Does communication involve learning (e.g. learning by being told)? Is any overall improvement of a multiagent system performance some kind of learning? An answer to these and related questions will require some more years of theoretical and experimental work in MAS learning. So instead of trying to answer these questions now we will point out the relationship between cooperation and learning.

First of all we may ask two negative questions: Why is there at all a need to cooperate? And why is there a need to learn? The answer to the second question is rather obvious: some agent needs to learn whenever it lacks some knowledge to perform some task— "perfect" knowledge has no room for learning. Learning has to do with improvement according to some criteria—i. e. amending those lacks for the task at hand. We can think the answer to the first question along the same line of thought: an agent needs to cooperate with other agents because it lacks some knowledge or some capability to perform a task. An agent with "perfect" knowledge and "complete" capabilities for a given task has no need to require the cooperation of other agents.

The parallelism of learning and cooperation stems from the fact that both are ways to deal with a agent's real shortcomings and lacks. We can summarize this parallelism as follows:

Learning Why is there a need to learn?
- Improving individual performance
- Improving precision (or quality of solutions)
- Improving efficiency (or speed of finding solutions)
- Improving the scope of solvable problems

Cooperation Why is there a need to cooperate?
- Improving individual performance
- Improving quality of solutions
- Improving efficiency in achieving solutions
- Achieving tasks that could not be solved in isolation

We are interested in investigating the interplay of learning and cooperation in this view. The approach presented in this paper explores a simple interplay of both: agents require the help of other agents when they are not capable of resolving a problem. In the future we hope to explore more complex interplays, for instance, when an agent can decide not to improve itself (not to learn) in situations when there is already a proficient agent in the MAS because it can simply require the help of this cooperative partner. The next subsection explains in more more details the approach we take.

1.2 Federated Peer Learning

We are investigating possible modes of cooperation among homogeneous agents with learning capabilities. Specifically, in this paper we are interested in a cooperative setting that assumes coordination among agents fulfilling the following conditions:

Homogeneous Agents The representation languages of the involved agents are the same. Consequently, communication among agents do not require a translation phase.

Peer Agents The involved agents are capable of solving the task at hand. In other words, cooperating agents are not merely specialists at specific sub-tasks. Instead, they are capable to solve the overall task by themselves (most of time, at least). This condition implies a peer to peer communication form.

Learning Agents The agents solve the task based knowledge acquired by learning from their individual, usually divergent, experience in solving problems and cooperating with other agents in solving problems.

We will call these conditions of agent cooperation a *federated peer learning* (FPL) framework. The FPL framework define a class of cooperative settings where learning can prove to have a clear leverage. In fact, we are focusing on the issue of how learning agents, that may have either the same method or several different methods for solving a given task and that moreover may can achieve a cooperative problem solving behavior that improves the individual behavior. The problem solving behavior of the agents will be biased by their individual learning based on their separate experience—since different sets of problems will actually occur in different locations. Consequently, even agents in principle similar can diverge as result of the individual learning experience, and cooperation may profit from these biasing by improving the overall performance of the involved agents.

In the FPL framework, we will focus in this paper on two modes of cooperation among case-based reasoning (CBR) agents. A CBR agent uses a form of *lazy learning* where past experiences are "generalized" (so to speak) by means of a similarity estimate between the current problem C and the precedent cases CB solved by the agent. The similarity-based reasoning (or analogical reasoning) involved follows the basic heuristic stating that *the more similar a case C is to a precedent $P \in CB$ the more similar the solution of C is to the solution of P*. While in eager forms of learning—like inductive techniques—the general descriptions for classes of solutions are built beforehand, lazy learning works in an on-demand, case-by-case basis. Learning in CBR can be seen as enlarging by means of a similarity estimate—thus, generalizing—a precedent case P until it includes the current case C [10]. We will show that the lazy nature of learning in CBR is very amenable to take advantage of cooperation.

The approach taken to communicate CBR systems is to extend Noos, a representation language developed at our Institute for integrating learning and problem solving that has been used to build several CBR systems [4]. The extension of Noos, *Plural* Noos, allows communication and mobile (or "migrating") methods among agents that use Noos as representation language. In particular, we will show two modes of cooperation among CBR agents: Distributed Case-based Reasoning (DistCBR) and Collective Case-based Reasoning (ColCBR). Intuitively, in DistCBR cooperation mode an agent A_i *delegates its authority* to another peer agent A_j to solve a problem —for instance when A_i is unable to solve it adequately. In contrast, ColCBR cooperation mode *maintains the authority* of the originating agent: an agent A_i can transmit a mobile method to another agent A_j to be executed there. That is to say, A_i *uses the experience* accumulated by other peer agents while maintaining the control on *how* the problem is solved.

Before explaining both DistCBR and ColCBR modes of cooperation in more detail, we will first introduce the task domain in which we are working.

1.3   The Task of Protein Purification

We have developed CHROMA, a system implemented in **Noos** that recommends chromatography techniques to purify proteins from tissues and cultures [5]. CHROMA includes two learning methods (a case-based method and an inductive method) and two problem solving methods (a CBR method and a classification method that uses the induced knowledge). Moreover, a metalevel method is able to prefer, for a particular problem, which problem solving method is more likely to succeed. Currently, we are simplifying the system for the cooperative CBR experiments and we will assume that CBR agents for protein purification will only embody one CBR method (see § 5 for future work on more complex situations).

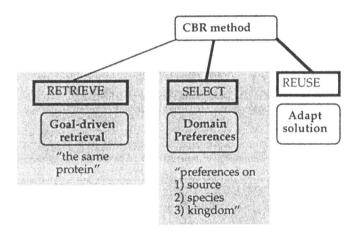

Fig. 1. The case-based reasoning method in CHROMA. The shaded part will be modified to adapt this method to a multiagent system (see Figure 7).

Why choose this task domain? The protein purification task is amenable to cooperative solutions since there are thousands of proteins and chromatography techniques are in current use in hundreds of industrial chemical labs that have their own bias as to the kinds of problems they regularly solve and the problems they seldom attack—but that can be regularly solved at another location. Moreover, different locations may have different methods for case-based reasoning that rely on a knowledge modeling analysis of their particular problems and their local expertise and biases.

The structure of the paper is as follows: first the **Noos** representation language is introduced and then the *Plural* **Noos** extension is summarized. Next, Distributed Case-based Reasoning (DistCBR) and Collective Case-based Reasoning (ColCBR) are discussed and their support by *Plural* **Noos** is explained.

Finally, some discussion about the generality of the approach and future work closes the paper.

2 Representation and Communication

The approach taken to develop cooperative CBR is to extend Noos, a representation language for integrating learning and problem solving that has been used to develop several CBR systems. In this section we first present some basic notions of the language, and later the *Plural* extension that supports communication and cooperation among CBR agents using Noos.

2.1 The Noos Representation Language

Noos is a reflective object-centered representation language designed to support knowledge modeling of problem solving and learning [3, 4]. Noos is based on the task/method decomposition principle and the analysis of knowledge requirements for methods —and it is related to knowledge modeling frameworks like KADS [15] or components of expertise [14][1]. A *method* models a way to solve a task. A method can be elementary or can be decomposed in subtasks. These new (sub)tasks can be achieved by corresponding methods in the same way. For a given task there may be multiple alternative methods (alternative ways to solve the task).

For instance, a CBR method [1] is decomposed into the **retrieve**, **select** and **reuse** subtasks and there are several possible methods to achieve each subtask. Decision-taking in Noos is modeled by a preference language that allows the specification of the conditions in which an alternative is better than others. Reasoning about preferences permits an agent to select a method from a set of alternatives or to choose to cooperate with an agent from a set of associate agents—as will be shown later.

The integration of learning and problem solving methods in Noos has two aspects. First, whenever some knowledge required by a problem solving method is not directly available there is an opportunity for learning. Secondly, learning methods are methods with introspection capabilities that can be analyzed also by means of a task/method decomposition. The basis for integrating learning methods is the *episodic memory*. The episodic memory stores the decisions taken during the inference—like successful methods engaged to tasks, results obtained by achieved tasks, and methods that have failed to achieve tasks. Noos provides two ways to perform introspection: using metalevel methods or using a set of retrieval methods provided by the language. Retrieval methods allow Noos to inspect and analyse previous specific situations in the episodic memory. For instance, case-based reasoning methods require to access stored cases, select one of them according to some criteria, and finally reuse the solution. The reuse task

---

[1] For related approaches see the Knowledge Engineering Methods and Languages web page at ftp://swi.psy.uva.nl/pub/keml/keml.html

reinstantiates the solution to the current problem or constructs a new solution according the precedent solution and the current problem[2].

An example of a case-based reasoning method used by CHROMA is the analogy-by-determination method. This method has a retrieve subtask with a retrieve-by-determination method that uses protein as determination[13]. This method retrieves from the episodic memory the solved experiments that satisfy the determination—purifying the same protein as the current experiment. The next subtask selects the most relevant precedent case according to domain knowledge criteria—like the kind of sample from which the protein is purified from. Finally, the last subtask reuse reinstantiates the purification plan of the most relevant precedent to the current problem. The knowledge required in this domain includes knowledge about proteins, chromatography techniques and purification plans.

Noos is an object-centered representation language based on *feature terms*. *Feature terms* are record-like data structures embodying a collection of *features*. Intuitively, a feature term is a syntactic expression that denotes sets of elements in some appropriate domain of interpretation. In this way feature terms can be viewed also as partial descriptions. The values of features are constants or other feature terms. Our approach is close to the ψ-*term* [2, 8] and *extensible records* [7, 9] formalisms.

The difference between feature terms and first order terms is the following: a first order term, e. g. $f(x, g(x, y), z)$, can be formally described as a tree and a fixed tree traversal order—in other words, variables are identified by position. The intuition behind a feature term is that it can be described as a labeled graph—in other words, variables are identified by name (regardless of order or position). This difference allows to represent partial knowledge.

Formally, we describe the Noos signature Σ as the tuple $\langle S, M, F, \leq \rangle$ such that:

- S is a set of *sort symbols* including $\bot, \top$;
- M is a set of *method symbols*;
- F is a set of *feature symbols*;
- $\leq$ is a decidable partial order on S such that $\bot$ is the least element and $\top$ is the greatest element.

Given the signature Σ and a set V of variables, we define a feature term ψ as an expression of the form:

$$\psi \quad ::= \quad X : s\,[f_1 \doteq \Psi_1 \cdots f_n \doteq \Psi_n]$$

where X is a variable in V, s is a sort in S, $f_1, \cdots, f_n$ are features in F, $n \geq 0$, and each Ψ_i is either a feature term, a set of feature terms or a method application $\#m$.

---

[2] In this paper we are focusing only in CBR learning methods—other learning methods like inductive methods [5] and analytical methods have also been integrated in this way.

Domain knowledge is represented in **Noos** by a collection of feature terms describing the concepts and their relations for a given domain. Feature terms have a correspondence to labeled graphs representation as shown in the description of an experiment in the chromatography domain of Figure 2.

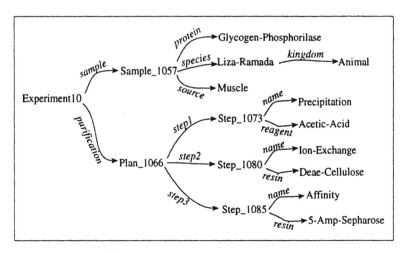

Fig. 2. A case description in CHROMA.

Methods are also represented as feature terms. The features of a method description represent the *subtasks* in which that method is decomposed. Methods are defined by refinement from a set of built-in methods. That is to say, a method is a feature term

$$\psi_m \quad ::= \quad X : m \, [f_1 \doteq \Psi_1 \cdots f_n \doteq \Psi_n]$$

as above except that now m is a sort in $\mathcal{M}$, i. e. it is a refinement of a built-in method.

The set of built-in methods in **Noos** are those of a general-purpose language plus some constructs enabling introspection. The uniform representation of methods as feature terms is what allows *Plural* **Noos** to transmit over the network both domain knowledge and methods in the same way.

Inference in **Noos** is on demand and is engaged by queries. For instance, solving the chromatography problem `experiment10` is engaged by querying the feature `purification` as follows: (`>> purification of experiment10`). The purification task is solved by the corresponding method associated with the `purification` feature of the problem. In the CHROMA system this method is the `analogy-by-determination` method explained below.

2.2 CBR in Protein Purification

We will introduce the CBR method used in our example domain of protein purification. We have to remark that **Noos** is not a CBR shell with a built-in, fixed way of performing case-based reasoning. Noos allows the *configuration* of a CBR system after a knowledge model analysis of the domain has been performed. Such a configuration is done with the component blocks provided by Noos–like generic retrieval methods—that are refined (or biased) in order to incorporate the domain knowledge we have modeled. In CHROMA the domain knowledge is used to characterize which features are more important when judging the similarity between a current problem and a precedent case. Noos allows to express such a knowledge by means of retrieval methods and preference methods. This abstraction permits to ignore implementation details like the indexing algorithms and, most importantly, will permit the communication of such methods among CBR agents. In this way a CBR agent can profit by lazy learning not only from the cases in its own Case-Base but also those cases known by other agents.

The configuration of the specific CBR method used in CHROMA is the following.

Goal-driven Retrieval The `retrieval` method is a generic method that selects from memory all cases obeying a constraint declared as `pattern`. Intuitively, it retrieves all cases subsumed by (all cases that match) the pattern. Domain knowledge in CHROMA state that we are interested only in cases where the `protein` feature has the same value as our current problem—and the rest of cases should be dismissed as irrelevant. This form of retrieval is called goal-driven retrieval (since the protein is the goal in our process) and can be represented by a general method called `retrieve-by-determination`.

Domain Selection Criteria A second component is a `preference` method that allows to impose a partial order among retrieved cases. In CHROMA there are three basic preferences:

Preference n. 1 Domain knowledge in CHROMA state that usually the most important criterion for similarity is having the same value in the `source` feature as in the current problem. This preference method imposes a partial order from the retrieved cases with that value to the retrieved cases that do not.

Preference n. 2 Another preference method is regarding the `species` feature—i.e. the species of the sample tissue or culture (`source`) from which the protein is purified. This preference discriminates the retrieved cases that are incomparable with preference n. 1.

Preference n. 3 The final component is also a preference regarding the `kingdom` taxon of the source, and it is applied to all retrieved cases that are not preferred among them by the preceding preference methods.

In our extension of CHROMA to distributed agents, each lab will supplement these general preferences with other specific preference criteria due to the kinds of problems they regularly solve and their local expertise. For instance, for a given tissues the specie criterion could be more relevant than the source criterion. Thus, each CBR agent will possibly contain selection criteria adapted to its own experience.

Reuse Finally, the last **reuse** method reinstantiates the purification plan of the most relevant precedent according to the previous domain preferences.

Learning in CBR is lazy: a CBR system imposes a partial order among (a relevant subset of) the past examples based on the current problem. The solution of a problem is determined by the solution of the case(s) that is maximal in the partial ordering established by preferences. Thus, solutions proposed by the system are function of the individual experience of the CBR system plus the domain knowledge given by the system designers during the knowledge modeling stage. Later in the paper we show how lazy learning plus method configuration can be used to support cooperation modes that improve the performance of a collectivity of CBR agents.

3 Agent Communication with **Plural Noos**

Plural provides a seamless extension of **Noos** that supports distributed scope and reference for all the basic constructs in **Noos**. A *Plural* **Noos** agent is a particular **Noos** application with a known address and with several *acquaintances*. An agent address is composed of one IP address, a port number and one identifier. The last identifier is needed since more than one agent can coexist within the same *Plural* **Noos** process. The acquaintances of an agent are those agents whose address is known by the agent —as in the actors model. Each *Plural* agent can have different acquaintances. If an agent A_i belongs to the acquaintances of an agent A_j, then A_j also belongs to the acquaintances of A_i.

A *Plural* **Noos** agent can be involved in solving only one problem at a time. Each problem solving process has a different identifier. When a *Plural* agent is solving a problem only accepts requests related to the same process identifier. In this way, possible deadlocks are avoided. Other deadlocks caused by circularities inside the same problem solving process are detected by the *Plural* **Noos** implementation. When an agent A_i requires a service from another agent A_j, and this one is already busy solving another different problem, A_i receives a *busy* message. Then A_i decides to wait some time to request the service to A_j again or ask it to another member of its acquaintances.

All *Plural* **Noos** agents taking part in an specific domain application share the same signature Σ. That is to say, the feature symbols, the sort symbols, and the method symbols are shared among all *Plural* agents involved in an application. So *Plural* **Noos** allows arbitrary **Noos** terms to be exchanged among one agent and its acquaintances. In particular, cases and CBR methods are terms that can be transmitted from a CBR agent to another. The CBR cooperation modes which this paper describes will use three *Plural* **Noos** capabilities: alien references, foreign evaluation, and mobile methods.

3.1 Alien References

Alien references extend **Noos** references to agents over the net. For instance, when the term identifier `experiment10` in `agent-i` is transmitted to `agent-j`, it is handled as an alien reference and it becomes naturalized as `experiment10@agent-i` by the `agent-j`. In the same way, a reference to a feature in `agent-i`, as (>> `purification of experiment10`), once transmitted to `agent-j` becomes an alien reference, (>> `purification of experiment10@agent-i`), in `agent-j`. Notice that feature symbols are shared among *Plural* Noos agents. If the value referenced by an alien reference in `agent-j` is needed then a transmission is automatically engaged asking for the value to `agent-i`. `Agent-i` is responsible for inferring that value and transmit it as answer to the `agent-j` request. Alien references avoid the problem of maintaining state when terms with state are transmitted over the network. State is local to agents and when an agent makes reference to a term which belongs to another agent, a alien reference is established [3]. Alien references are transmitted over the network: if `experiment10@agent-i` is a value of the feature `purification` of entity `experiment21` in `agent-j` and a new agent `agent-k` has the reference (>> `purification of experiment21@agent-j`) eventually `agent-k` will get the alien reference `experiment10@agent-i`.

Alien references make up the basic mechanism that underpins the exchange of terms among *Plural* Noos agents over the network. In essence, as Noos terms can be seen as labeled graphs, and since Noos performs a lazy evaluation, not all the nodes in a graph are transmitted when the root is referenced by a remote agent. Instead, the transmission of a term from an agent `agent-i` to another agent `agent-j` starts by sending the graph root (an identifier, the sort, and the name of the root features). If the graph node sent to `agent-j` is a constant (a number, a string or a sort) a local reference is established by `agent-j`. Otherwise, `agent-j` establishes an alien reference to that node. When `agent-j` requires the value of any of the features of that node, a new transmission is engaged asking for it to `agent-i`. Then `agent-i` inferres its value and sends it to `agent-j`. Path equality (sharing) and circularities in the graph are preserved.

The next example describes how the term `experiment10` (see Figure 2) in `agent-i` is transmitted to `agent-j`. In the first step the term identifier `experiment10` and the names of its features `sample` and `purification` are sent to `agent-j`. Since `experiment10` is not a constant, an alien reference will be established in `agent-j`, as showed in Figure 3. Then, if the value of the feature `sample` is required by `agent-j`, it will be automatically requested to `agent-i`. Next, `agent-i` will resolve that reference to `sample_1057`. This term identifier and the names of its features `protein`, `species` and `source` will be sent back to `agent-j`, and a new alien reference `sample_1057@agent-i` will be established (see Figure 4) in `agent-j`, since `sample_1057` is not a constant. Figure 5 shows the state achieved once the values of features `protein`, `species` and `source` have been required by `agent-j`. Values `Glycogen-Phosphorilase`, `Liza-Ramada` and `Muscle` are all of them sorts, and since sorts are shared, a reference to the local sorts has been established in `agent-j`, when they have been received. Finally, as Figure 6 shows, when the value of feature `kingdom of Liza-Ramada` is re-

---

[3] Our approach is similar to that of the distributed object-oriented language Obliq [6] regarding the fact that alien references are local to a site (here, an agent). A major difference is that *Plural* transmits *terms* over the net while Obliq transmits *closures*.

quired by `agent-j`, the value `Animal` is inferred in `agent-j`, without need to ask `agent-i`, since `Liza-Ramada` is local to `agent-i`.

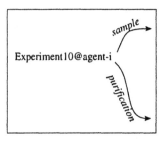

Fig. 3. An alien reference to `experiment 10` at `agent-i` is established in `agent-j`.

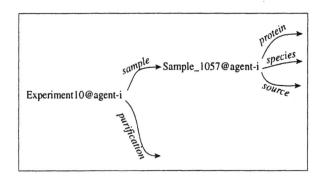

Fig. 4. An alien reference to `Sample_1057` at `agent-i` is established in `agent-j`.

3.2 Foreign Evaluation

The *Plural* Noos foreign evaluation capability allows an agent to use a method owned by another agent —as in remote procedure call (RPC). Specifically, foreign evaluation allows an agent `agent-i` to ask another agent `agent-j` to execute a specific method using the parameters given by `agent-i`, as in the next expression of `agent-i`.

```
(define (foreign-eval)
  (method (define (protein-purif-method)
          (case experiment10)))
  (at agent-j))
```

In this expression, an agent `agent-i` asks to another agent `agent-j` to evaluate the method `protein-purif-method` using as `case` the `experiment 10`. Then

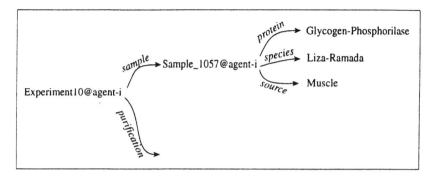

Fig. 5. Glycogen-Phosphorilase, Liza-Ramada and Muscle are references to sorts. Since sorts are shared by all agents they do not require alien references.

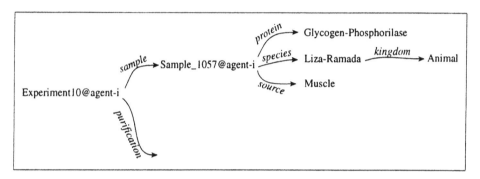

Fig. 6. The value of feature kingdom is inferred from the local sort.

agent-j will start the evaluation of its own **protein-purif-method** method, annotating that this evaluation is being performed for the remote agent **agent-i**. When the **case** feature of this method is required during the evaluation, it will be automatically asked to **agent-i**. Then the **experiment10** term will be sent from **agent-i** to **agent-j**, such as was explained in the last subsection. This value will be an alien reference in **agent-j** and will become naturalized as **experiment10@agent-i**. During the evaluation, further references in **agent-j** to features of **experiment10@agent-i** are interpreted as alien references as well. And its values will be transmitted from **agent-i** as they are needed. Once **agent-j** finishes the evaluation of method **protein-purif-method** the result got will be sent back to **agent-i**, as answer to the evaluation of the **foreign-eval** method.

3.3 Mobile Methods

For some cooperation modes it is necessary to support so-called mobile (or migrating) methods. In *Plural* Noos a mobile method defined in an agent **agent-i** can be transmitted to any member of its acquaintances. When an agent **agent-i** sends a mobile m thod to **agent-j**, this process involves also transmitting the whole task/method decomposition to `agent-j`—i. e. the subtasks of that method, and the methods for those subtasks. The process of sending a mobile method, called `jump`, consists of

1. sending the the name of the built-in of which the method is a refinement
2. the names of its features (i. e. the method's subtasks)
3. Recursively, the methods defined for those subtasks

While foreign evaluation requires the remote agent to own a particular method which can be used by the originating agent, the mobile methods capability of *Plural* Noos does not require it.

Mobile methods are supported by the *Plural* Noos capability of transmitting method terms. A mobile method term is first defined in an originating agent `agent-i`:

```
(define (jump)
  (method (define (mobile-method-k)
          (description (>> description of problem-13))))
  (at agent-j))
```

When a method jumps to a remote agent, the whole task/method decomposition of the mobile method is transmitted in a lazy way similar to that explained in § 3.1. Nevertheless, there is a main difference between the `jump` process and the transmission of a feature term resulting from an alien reference. In the `jump` process, when a reference is made to a feature of a mobile method (i. e. a subtask), *Plural* Noos requests to the originating agent the *method* corresponding to that feature name. In this way, the whole task/method decomposition of the mobile method will be transmitted, on demand, from the originating agent to the target agent.

4 Modes of Cooperation for CBR Agents

Since learning is lazy in CBR systems, cooperation involves expanding the set of precedents to be used in similarity-based reasoning from the individual memory of a CBR agent to the memories of a collectivity of CBR agents. We argue that there are two general ways to do so: Distributed Case-based Reasoning (DistCBR) and Collective Case-based Reasoning (ColCBR). Intuitively, both DistCBR and ColCBR are based on solving a problem by reusing with the knowledge learned by other CBR agents. Given an agent (the *originator*) trying to solve a given problem, the difference between both modes is regarding which similarity-based reasoning method is used: that of the originator or that of the CBR agent that is helping the originator.

In other words, the difference is the following:

DistCBR is based on an agent transmitting the problem and the task to be achieved to another agent, and the CBR method used is that of the receiving agent. In this sense, the CBR process is *distributed* since every agent works using its own method of solving problems.

ColCBR is based on an agent transmitting also the method that is to be used to solve that problem to another agent (and that method will use the knowledge learnt by the receiving agent). In other terms, the originator is using the memory of the other agents as an extension of its own—as a *collective memory*—by means of being able to impose to other agents the use of the CBR method of the originator.

From the standpoint of implementing those cooperation modes, we can say that DistCBR is supported by the foreign evaluation capability and ColCBR is supported by mobile methods (also called "remote programming") capability of *Plural* Noos.

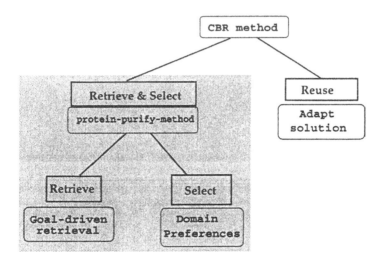

Fig. 7. The case-based reasoning method for DistCBR and ColCBR in CHROMA. The shaded part is changed from that of Figure 1 and are the subtasks performed by other agents on request of the originating agent.

Regarding the chromatography domain, the CBR method for CHROMA shown in Figure 1 is modified as shown in Figure 7. Since the shaded part in both figures is the part that an originating agent wants to ask other agents to perform over their own case-bases, we introduce a new method, **protein-purify-method**, that simply gathers together both tasks, **retrieve** and **select**, that have to be distributed over other agents. In this way, DistCBR will be implemented using **protein-purify-method** by foreign evaluation and ColCBR will be implemented using **protein-purify-method** as a mobile method.

4.1 Distributed Case-based Reasoning

The DistCBR cooperation mode is, intuitively, a class of cooperation protocols where a CBR agent A_{orig} is able to ask to one or several other CBR agents $\{A_1...A_n\}$ to solve a problem on its behalf. The cooperation mode definition leaves to specific protocols designed for given task domains the specification of which criteria an agent A_{orig} uses to ask another to solve a problem, how to choose which agents to ask and in which order. DistCBR is based on the *Plural* Noos capability of *foreign evaluation*. A specific protocol for the protein purification task is given below. This is not a shortcoming or underspecification of our framework: since these issues and decisions are domain-dependent they are to be established by a knowledge modeling analysis of the task domain that later implemented by Noos methods. The only difference is that these *Plural* Noos methods will have references to —and will engage communication with— other agents. ·

DistCBR involves two main cooperation tasks: a) A_{orig} sends the (identification of the) current case C_{curr} to an agent A_j, and b) asking A_j to solve the purification task on the case C_{curr}. As result, agent A_{orig} receives a solution inferred by A_j based on its own $CBR - method_j$ and its case-base CB_j —or a failure token. Upon a failure of the agent A_j, A_{orig} can iterate the cooperation tasks with the next agent of its preference.

An agent in DistCBR CHROMA has a set of acquaintances $\{A_1...A_n\}$ that are agents having at least a CBR method for solving protein purification problems and a case-base of such problems already solved. A_{orig} can prefer to ask first to an agent A_i that has previously solved for it a problem regarding the same protein (goal preference)[4]. In general, each CBR agent may have a different protocol for deciding which agent to ask to solve the current problem.

In order to start a DistCBR cooperation, the originating agent only needs to know the name (identifier) of the CBR method used by each acquaintance for the task `purification`—by convention we will assume all agents use the same public name `protein-purif-method`[5].

The cooperation tasks of DistCBR are achieved in the implementation by requiring the `foreign-eval` of a M_k (say `protein-purif-method-k`), for each RM_k in the collection of methods for the `retrieve-&-select` task. The *Plural* Noos syntax is as follows:

```
(define (foreign-eval)
   (method (define (protein-purif-method-k)
           (case case-33)))
   (at agent-j))
```

---

[4] This is the same preference that the stand-alone CHROMA system applies in the retrieval task (prefer a case with the same protein as the current problem).

[5] These method names can be easily acquired asking the acquaintances (`>> method of (task purification of purification-problem at agent-j)`) but we have no room for the discussion here.

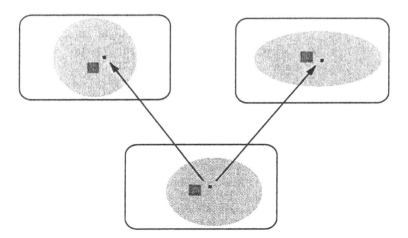

Fig. 8. In Distributed CBR each agent uses its own `retrieve-&-select` method on the current problem. The shaded areas represent a similarity degree centered around the current problem (the black dot). The most similar case in each agent's memory is depicted as a shaded box. The shaded areas are in general different because the criteria that specify what is "similar" may vary from one agent to another. Compare to Figure 9 that shows the effect of using a unique mobile CBR method in ColCBR.

This process can be iterated on other acquaintances until a solution can be obtained for an agent that has an appropriate case precedent for the current problem.

The current implementation of DistCBR CROMA has two strategies to select the acquaintances to which an agent asks help. The first one, as mentioned, simply asks other acquaintances in some specific order until one of them can solve the problem requested using its own method. The results obtained in its strategy for DistCBR cooperation mode crucially depends on to ordering in which an agent selects an acquaintance, and a more complex handling of it is discussed at § 5. A second strategy, that we call *conservative*, allows more control to the originating agent at the cost of more communication. The conservative strategy of DistCBR involves the originating agent asking to solve the problem to all its acquaintances and obtaining the best cases according to them. Then, the originating agent can select with of them is best according to its own criteria, for instance according to the Preferences in § 2.2.

4.2 Collective Case-based Reasoning

The ColCBR cooperation mode is, intuitively, a class of cooperation protocols where a CBR agent A_i is able to send a specific CBR method $CBR - method_i$ of its choosing to one or several CBR agents $\{A_1...A_n\}$ that are capable of using that method with their case-base to solve the task at hand. ColCBR is based on the *Plural* Noos capability of *mobile methods*: an originating agent A_i can define a method $CBR - method_i$, bind it to the current problem C_{curr}, and migrate it to another agent A_j that has previously solved for it a problem regarding the same protein (goal preference). The mobile CBR method, upon transmission to A_j, can perform the CBR subtasks (**retrieve**, **select**, **reuse**) using the case-base CB_j. When the mobile CBR method finishes the result (or a failure token) is sent back to A_i. The originating agent A_i can then decide if it is necessary to send the mobile CBR method to a new acquaintance and start a new iteration.

In the chromatography domain, the cooperation tasks of ColCBR are achieved as follows. First, a CBR method for protein purification **cbr-pp-mobile-method** is defined in originating **agent-i**; then the method is bounded to the current problem **case-33** and sent to **agent-j** by the expression:

```
(define (jump)
  (method (define (cbr-pp-mobile-method)
             (case case-33))
  (at agent-j))
```

This is equivalent to the following process:

1. The identifier of **cbr-pp-mobile-method** is sent to **agent-j**,
2. Since the method is defined in **agent-i**, **agent-j** requests the subtasks of **cbr-pp-mobile-method**; as result **agent-j** will receive the methods for those subtasks and (the identifier of) **case-33**.
3. Recursively, the methods of the subtasks will be transmitted and their subtasks methods will be requested, until all the task/method decomposition is transmitted to **agent-j**.
4. Finally, **cbr-pp-mobile-method** is executed by **agent-j** and the result is returned to the originating **agent-i**.

In general, the originating agent in ColCBR can have *several* mobile methods for a task. In ColCBR an agent could have several mobile CBR methods with a preference ordering among them from the more constrained CBR method to the less constrained. In this way, the agent can assure that it can retrieve the precedent cases from the distributed case-base that comply to the most relevant requirements for the task, and only when no precedent is found, a second mobile CBR agents searches for a less relevant precedent case in the distributed case-base.

In the current implementation of ColCBR CHROMA the agents follow *conservative* strategy in asking for help to other agents the rationale of which is to assure a result as close as possible to the original CBR method for a standalone system. In particular, ColCBR CHROMA *conservative* strategy tries to find the best precedent case known by a federation of agents (its acquaintances)—where "best" is interpreted in the sense of the preferences explained in § 2.2.

In order to do so, an agent with this strategy has a collection of methods $\{M_1, M_1^m, \ldots, M_4, M_4^m\}$ for the task **retrieve-&-select**. The first two M_1 and M_1^m are methods that considers the preferences in § 2.2 as restrictions: in this way it can retrieve only the precedent cases satisfying all these conditions.

Method M_1 retrieves cases from the case-base of the originating method. If this method fails—i. e. there is no such a case in memory—Noos backtracks taking the second option, namely M_1^m, that is a mobile method version of method M_1. M_1^m is sent one by one to all the acquaintance—if the mobile method sent to an agent fails, *Plural* Noos sends the mobile method to the next acquaintance. If one of them returns such a case the **retrieve-&-select** task is finished. Otherwise it means that all agents have failed—none of them have a precedent case satisfying all the constraints in § 2.2. In this situation, Noos selects the next method, namely M_2. Now both M_2 and M_2^m are a less restricted version of M_1 and M_1^m where the less important constraint in § 2.2 (Preference 3) is dropped. Using M_2 and M_2^m now DistCBR CHROMA can retrieve a precedent case from its memory or one of its acquaintance receiving the mobile method M_2^m can retrieve a precedent case from its memory. Again, if any mobile method retrieves a case complying to the constraints the process stops, otherwise proceeds with M_3 and M_3^m (that only requires as constraint Preference 1 in § 2.2) and with M_4 and M_4^m (that only performs Goal-driven Retrieval but enforces no preference).

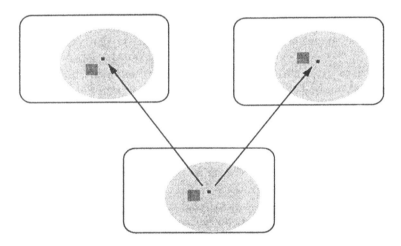

Fig. 9. In Collective CBR mobile methods assure that the similarity considered will be the same in all agents. The shaded area represents a similarity degree centered around the current problem (the black dot). The most similar case in each agent's memory is depicted as a shaded box. The shaded area is equal in the originating agent and in the two agents that receive a mobile CBR method, while in DistCBR (see Figure 8) they are different.

It is easy to see that this strategy assures that the originating agent finds the most preferred case according to the established preferences from any case base of an acquaintance agent. Figure 8 shows intuitively the effect of a mobile CBR method: the same retrieval and selection method is used in each agent, the only difference being the case that is retrieved in each agent according to its past experience.

5 Future Work on Cooperative Case-based Reasoning

The conservative strategy of last section is not obliged by neither by the ColCBR mode of cooperation nor by the *Plural* Noos language. It is perfectly possible and rational that the originating agent sends a mobile CBR method that embodies the preferences in § 2.2 to the acquaintance agents. In this strategy, the first acquaintance agent that has case satisfying *some* of the preferences in § 2.2 will be retrieved. In this strategy, the order in which we take the acquaintance agents to solicit them to solve a problem becomes crucial. There are two approaches to solve this issue: instituting authority and learning competence models. *Instituting authority* involves selecting a priori the class of problems for which each agent is competent on and giving him authority to solve them. This selection can be typically established by the designer of a multiagent system (MAS) or by the institution(s) that grant the cooperation of one of its agent into a MAS. The second approach involves agents learning a model a *competence model* of other agents in a MAS—i. e. each agent has to determine (learn) an individual model of which problems other agents in a MAS are competent to solve. This approach is high in our research agenda on federated learning.

Although the CBR cooperation modes we propose are quite general descriptions, there are more options that those explained in this paper and that are envisioned as future work. For instance, we plan use the full CHROMA application which integrates induction and CBR. In this setting, DistCBR would use the metalevel method of CHROMA that selects the appropriate problem-solving method; while ColCBR the originating agent would be able to send to other agents the method of its choosing.

A variant of the DistCBR and ColCBR cooperation modes consists of asking k acquaintances to solve the problem instead of asking one by one until a solution is achieved. This variant requires a new task on the originating agent that performs some selection of the solution or consensus aggregation function. Both selection and consensus require A_{orig} having a model of the reliability of the agents involved —the model can be based on some learning method based on the previous results of those agents. However, the selection and consensus processes do not pertain to the cooperation mode as such, but to the knowledge modeling analysis of the task domain. For instance, in our domain more than one chromatography plan can effectively purify a protein, so it is possible to recommend more than one correct solution (although a solution ranking is of course highly desirable).

6 Discussion

We have presented two simple yet powerful cooperative modes of case-based reasoning and learning. Even assuming that all the participating agents start with the same CBR method, the individuality of the learning agents (the separate existence of agents having different memories given by disparate past experience) implies a distinct content (resulting from learning) for each agent. In the

DistCBR cooperation mode an originating agent *delegates authority* to another peer agent to solve the problem. In contrast, ColCBR *maintains the authority* of the originating agent, since it decides which CBR method to apply and merely *uses the experience* accumulated by other peer agents.

In the protocols developed for the chromatography domain, since an agent only cooperates with another agent when the originator is not able to solve a problem (according to the domain knowledge constraints), the result of cooperation is always better than no cooperation, and communication is engaged only when need be. However, these protocols are domain dependent and are the result of a knowledge modeling process. The cooperation modes are, we argue, general for agents that capable of lazy learning.

The lazy nature of learning in CBR helps in the reuse and exploitation of the experience of different agents in a cooperative setting. Since the implicit generalization of similarity-based reasoning is performed on a case-by-case basis, and the cooperation is also made on a case-by-case basis, both can be integrated seamlessly. Eager learning, as induction, perform learning over sets of cases and built new knowledge structures capable of solving new problems—and some of them discard the particular cases after induction. In this setting the Distributed Mode seems applicable, since every agent uses the induced knowledge structures to solve a particular problem. However, the Collective Mode seems problematic— inapplicable in fact if the agents discard the particular cases. This mode is based on the idea of extending the memory of an agent to the memory of the rest of agents by forming a collective memory. However, the distribution of agents and experience can meaningfully exploit the collective memory in a lazy, on-demand way. An eager use of collective memory, for instance, would be for an agent to perform induction over *all* cases known to all associate agents. This option implies a communication overhead and in the long run amounts to a centralized view of learning where every agent is aware of all the accumulated experience of every other agent.

Related work is KQML and CBR-TEAM. The communication capabilities of *Plural* Noos are compatible to the basic constructs of KQML [11]. Since we are dealing with homogeneous peer agents the rather general features of KQML (like ontologies and representations translation) are not needed, there is no need for *Plural* to use the KQML equivalent constructs[6]. It remains future work to see if Noos agents communicating with agents using other representation languages like Loom or KIF could actually use KQML constructs. The CBR-TEAM system uses negotiated case retrieval as a form of cooperative CBR among heterogeneous agents (subtask specialists) [12]. The overall task is a distributed constraint optimization process over the shared interface parameters (parameters optimized by more than one agent).

In this paper we have focused on modes of cooperation among agents able to perform some lazy learning, but we focused the learning process on learning about the task domain—chromatography techniques in our application. How-

---

[6] An example of equivalence is the following. KQML has an `ask-all` construct. Finding all solutions of a task in Noos syntax is written (`*>> task of problem at agent`).

ever, as a result we are quite aware that learning has also to play a major role regarding the cooperation process itself. We plan to study this issue by the agents being capable to learn *competence models* of other agents. We think this approach can be useful for any MAS where the authority of an agent is not predetermined by the system designer. In fact, we can think about Federated Peer Learning as a framework in which the authority of each participating agent is dynamically allocated by other participant agents assessing their scope and degree of competence.

Acknowledgements

The research reported on this paper has been developed at the IIIA in the framework of the ANALOG Project (CICYT grant TIC 122/93),the SMASH Project (CICYT grant TIC 96-1038), a CSIC fellowship, and DGR-CIRIT fellowship FI-DT/96-8472.

Updated information will be posted on the WWW at URL http://www.iiia.csic.es/Projects/FedLearn/CoopCBR.html and http://www.iiia.csic.es/Projects/FedLearn/plural.html.

References

1. Agnar Aamodt and Enric Plaza. Case-based reasoning: Foundational issues, methodological variations, and system approaches. *Artificial Intelligence Communications*, 7(1):39–59, 1994. Available online at <url:http://www.iiia.csic.es /People/enric/AICom_ToC.html>.
2. Hassan Aït-Kaci and Andreas Podelski. Towards a meaning of LIFE. *J. Logic Programming*, 16:195–234, 1993.
3. Josep Lluís Arcos and Enric Plaza. Integration of learning into a knowledge modelling framework. In Luc Steels, Guss Schreiber, and Walter Van de Velde, editors, *A Future for Knowledge Acquisition*, number 867 in Lecture Notes in Artificial Intelligence, pages 355–373. Springer-Verlag, 1994.
4. Josep Lluís Arcos and Enric Plaza. Inference and reflection in the object-centered representation language Noos. *Journal of Future Generation Computer Systems*, 12:173–188, 1996.
5. Eva Armengol and Enric Plaza. Integrating induction in a case-based reasoner. In J. P. Haton, M. Keane, and M. Manago, editors, *Advances in Case-Based Reasoning*, number 984 in Lecture Notes in Artificial Intelligence, pages 3–17. Springer-Verlag, 1994.
6. Luca Cardelli. Obliq, a language with distributed scope. Technical report, DEC, Systems Research Center, 1995.
7. Luca Cardelli and John Mitchell. Operarions on records. In Carl A. Gunter and John C. Mitchell, editors, *Theoretical aspects of object-oriented programming: types, semantics and language design*, Foundations of computing series, pages 295–350. MIT Press, 1994.
8. B. Carpenter. *The Logic of typed Feature Structures*. Tracts in theoretical Computer Science. Cambridge University Press, Cambridge, UK, 1992.

9. Laurent Dami. *Software Composition: Towards an Integration of Functional and Object-Oriented Approaches.* PhD thesis, University of Geneva, 1994.

10. D. Dubois, F. Esteva, P. Garcia, L. Godo, and H. Prade. Similarity-based consequence relations. In Ch. Froidebaux and J. Kohlas, editors, *Symbolic and Qualitative Approaches to Reasoning and Uncertainty*, number 946 in Lecture Notes in Artificial Intelligence, pages 171–179. Springer-Verlag, 1995.

11. Tim Finin, Jay Weber, and et al. Specificastion of the kqml agent-communication language. Technical report, The DARPA Knowledge Sharing Initiative, 1994. <http://retriever.cs.umbc.edu/kqml/kqmlspec/spec.html>.

12. M V Nagendra Prassad, Victor R Lesser, and Susan Lander. Retrieval and reasoning in distributed case bases. Technical report, UMass Computer Science Department, 1995.

13. S. Russell. *The use of knowledge in Analogy and Induction.* Morgan Kaufmann, 1990.

14. Luc Steels. Components of expertise. *AI Magazine*, 11(2):28–49, 1990.

15. Bob Wielinga, Walter van de Velde, Guss Schreiber, and H. Akkermans. Towards a unification of knowledge modelling approaches. In J. M. David, J. P. Krivine, and R. Simmons, editors, *Second generation Expert Systems*, pages 299–335. Springer Verlag, 1993.

Contract-Net-Based Learning in a User-Adaptive Interface Agency

Britta Lenzmann and Ipke Wachsmuth

University of Bielefeld
Faculty of Technology
AG Knowledge-Based Systems
D-33501 Bielefeld, Germany
{britta,ipke}@techfak.uni-bielefeld.de

Abstract. This paper describes a multi-agent learning approach to adaptation to users' preferences realized by an interface agency. Using a contract-net-based negotiation technique, agents as contractors as well as managers negotiate with each other to pursue the overall goal of dynamic user adaptation. By learning from indirect user feedback, the adjustment of internal credit vectors and the assignment of contractors that gained maximal credit with respect to the user's current preferences, the preceding session, and current situational circumstances can be realized. In this way, user adaptation is achieved without accumulating explicit user models but by the use of implicit, distributed user models.

1 Introduction

Interface agents are computer programs that enhance the human-computer interaction by mediating a relationship between technical systems and users [Lau90]. On the one hand, they provide assistance to users by acting on his/her behalf and automating his/her actions [Nor94]. On the other hand, they allow more human-like communication forms by translating qualitative input of the human user to precise commands which can be interpreted by the application system [WC95].

To assist the user in performing tasks, interface agents need to have knowledge about the user and the application. A prominent approach is to build learning interface agents that automatically acquire knowledge about tasks and preferences of the user by applying machine learning techniques [Mae94]. In such approaches, a single personal interface agent is used which customizes to an individual user by acquiring explicit user data.

Since acquiring user-specific data and building explicit user models has found critique with respect to privacy of personal information [Nor94], we pursue a different approach where an *interface agency* – consisting of multiple sub-agencies – customizes to users' preferences by building an implicit, distributed model of the user. We use learning from indirect user feedback that allows to determine which agents of different sub-agencies are preferred by the individual user and in the actual situation. Internally, the dynamic activation of single agents is realized by a contract-net learning process where multiple agents in the role of contractors as well as agents in the role of managers negotiate with each other to pursue the common goal of dynamic user adaptation.

The paper is structured as follows: We start by distinguishing our approach from related work in the field of user adaptation and learning interface agents. We proceed by characterizing our learning technique in terms of the learning category and the form of learning. Having described the basic ideas and requirements, we focus on the realization of agents in the role of contractors as well as agents in the role of managers. The learning technique which has been integrated in the interactive 3D-graphics system VIENA is then illustrated by an example of adaptation to users' preferences for different spatial reference frames.

2 Related Work

To realize an automatic adaptation to the varying preferences different users could have for the possible solutions produced by an application system, a computer system must have knowledge about its users and the application domain. Conventional approaches in the fields of Artificial Intelligence (AI) and Human-Computer Interaction (HCI) are commonly based on the modelling of users where information about the users is gathered, represented, and used to adapt to users' requirements [McT93]. When acquired explicitly, a user model demands a large amount of work from the application designer (or the user in the case of end-user programming), and it is relatively static. In the case of automatic aquisition, such disadvantages do not occur. Nevertheless, both methods have the draw-back that knowledge about the user is kept in a explicit knowledge base; this fact has found critique with respect to privacy of personal information [Nor94].

Interface agents are considered as metaphor of indirect management [Kay90] instead of interacting via commands or direct manipulation. For example, [Chi90] realized a learning interface agent that helps a user solve problems in using the UNIX operating system. Such approaches to user adaptation are based on endowing the interface agent with a huge amount of background knowledge about the user and the application [MK93]. Similiar to the methods described above, they have the disadvantage that a huge amount of fixed knowledge has to be acquired.

More elaborated approaches to learning interface agents rely on the automatic aquisition of knowledge about tasks and preferences of the user by applying machine learning techniques. For example, [Mae94] uses the techniques of as learning from observations, learning from feedback, learning from examples, or learning from other agents to build learning interface agents for electronic mail handling and information filtering. [MCFMZ94] use decision trees to realize a learning personal assistant for meeting scheduling. [Sel94] has built a user-adaptive teaching agent that supports users in writing Lisp programs by watching the users actions and using a production system and a blackboard mechanism. The main difference to our approach, presented in this paper, is that a single personal interface agent is used which customizes to an individual user by acquiring user data. Thus, the problem of keeping personal information of users still consists.

Although a number of multi-agent learning techniques exists, relatively few ones focus on the aspect of user adaptation. In [LMM94], agents collaborate with each other to steepen the agents' learning curves and to handle unencountered situations. With respect to contract-net negotiation, the integration of learning facilities has been studied by some researchers. In [Dow95], for example, agents of a multi-agent system learn about other agents' abilities and task load to switch from broadcasting task announcements to direct task assignment. [OHA96] has realized the multi-robot learning system LEMMING that learns to determine the most suitable robot for task execution by using case-based reasoning within the contract-net negotiation. Similiar to our approach, useful information of previous negotiation messages is extracted. In contrast to our approach, their work is limited to the aspect of optimizing communication overhead. In summary, using contract-net-based learning to achieve user adaptation without building explicit user models seems to be an innovative approach.

3 Characterizing the Learning Method

Before describing the contract-net-based learning method in detail, we will characterize our approach in terms of the learning category and the form of learning with emphasis on the learning strategy and the learning feedback as proposed in [Wei96].

3.1 Learning Category

Allowing adaptation to users' preferences without building explicit user models relies on the design of a multi-agent system that consists of numerous sub-agencies realizing different functionality types. Each sub-agency joins agents of the same type but with slightly different internal functionalities together and corresponds to a class of users' preferences with respect to a functionality type. Each agent of a sub-agency realizes a specialized preference of its sub-agency's preference class.

By exchanging information via a contract-net negotiation mechanism, the agency, as a unit, learns by organizing itself in the way that those agents of each sub-agency can be activated which correspond best to the actual preferences of the current user in the given situation. This means that learning is aimed at a dynamic adaptation with respect to

1. preferences of different users, as well as
2. time-varying preferences of an individual user during a session.

In this way, the approach can be categoriezed as multi-agent learning insofar as it requires the presence of multiple agents which negotiate with each other to pursue the common goal of dynamic user adaptation.

3.2   Form of Learning

The agency learns the goal of user adaptation by offering solutions to the user and observing the user's feedback, i.e., implicit positive and explicit negative feedback. Implicit positive feedback is given when a user's instruction is followed by any instruction which does not decline the previous one. Explicit negative feedback is given when the user corrects the solution offered by the interface agency.

The feedback provided by the user acts as a critic which is interpreted and encoded by the interface agency in the form of credit values. Credits are stored locally by each agent and correspond to agents' strengths at discrete interaction steps. Until the user's instructions are evaluated entirely with respect to his/her preferences, a number of sub-tasks have to be solved by the interface agency. Therefore, a number of communication and cooperation processes are carried out between different agents within and across sub-agencies. The overall adaptation which emerges from consecutive feedback by the user is then achieved by the cooperation of agents.

The learning strategy decribed here can be classified as a form of *learning by discovery* since agents capture knowledge about the interaction process and about other agents by making obeservations and generating solutions on the basis of the observational results. From the perspective of the feedback that is available to the agency the method can be regarded as *reinforcement learning* since the agency has to learn the right actions by not precisely specified feedback.

4   Contract-Net-Based Learning

Considering contract-net negotiation within a multi-agent system, the execution of tasks and sub-tasks is the result of a bidding scheme between agents in the role as a contractor and agents in the role as a manager [DS83]. Once a task has to be executed, a manager agent sends a task announcement to its contractor agents. With respect to the task description, idle contractors generate bids and send them to the manager. The manager evaluates all incoming bids and selects one or more contractors for task execution.

4.1   Extending Standard Contract-Net Negotiation

On the basis of the standard contract-net mechanism, we have integrated the following two steps into the contract-net negotiation to achieve user adaptation:

1. Adjustment of credits in correspondence to the user feedback and the actual situation parameters
2. Assignment of those agents that are eligible for the task and have maximal credits in correspondence to the interaction process

These steps are realized by different agent instances of the interface agency. The first step is realized by agents currently in the contractor role which extract and store relevant information from incoming negotiation messages, consider actual situation parameters, use the results to adjust credits, and include current credits in their bids. The second step is realized by agents as managers which compare all incoming bids by evaluating credits and the negotiation history, allocate the task to the most promising contractor, and reject the bids of all other contractors. Figure 1 illustrates the negotiation process for the case of two contractors and one manager[1].

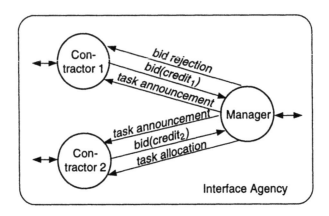

Fig. 1. A detail of the negotiation process: Contractor 2 generates the better bid and gets the task whereas the bid of the potential contractor 1 is rejected.

In this way, the overall adaptation to users' preferences is achieved by a set of negotiation functions that are executed locally between managers and contractors and results are distributed in the agent system. Each of these functions realizes the two steps mentioned above by the concatenation of the functionality of a manager and of a number of contractors. For the case of a manager k and m contractors, the negotiation function at an interaction step t is defined by:

$$f_{neg}(X(t)) = manager_k(contractor_1(X_1(t)), \ldots, contractor_m(X_m(t)))$$

Each contractor function $contractor_i$ realizes the adjustment of credits based on the data $X_i(t)$ extracted from the current and preceding negotiation messages. On the basis of these computations, the manager function $manager_k$ realizes the assignment of the most successful contractor. A schematic view of the negotiation function is shown in Figure 2.

Taking a more abstract point of view, the activation of preferred agents of different sub-agencies gives rise to an implicit, distributed user model where the acquisition of explicit user models is avoided. Besides using the contract-net

---

[1] We have adjusted terminology used in earlier publications to be consistent with [DS83].

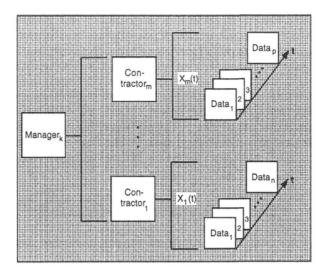

Fig. 2. Functional scheme of the contract-net-based learning approach

mechanism as tool to achieve a user-adaptive multi-agent system, extensions of the standard negotiation process as described at the beginning of this section concern:

- contractors and managers store knowledge about the preceding negotiation processes
- this knowledge is used, by contractors, to adjust credits and, by managers, to assign contractors
- bids include user feedback and credits corresponding to the actual situation

4.2 Requirements of the learning strategy

A prototype version of the adaptation method, where simple heuristics is used concerning the modification of credits and the assignment of contractors, is described in [LW96]. With this version, our system was already able to adapt to changing preferences of users. A disadvantage of this first prototype system is that credits are simply represented by scalar values that loose information about the history about potentially successful contractors and, thus, allow only very short-term adaptation. Moreover, the method is not appropriate for handling any given number of contractors and managers. Finally, no current situation parameters are taken into account when adjusting credits.

To allow a more flexible adaptation, our first version has been further elaborated with respect to the manager and contractor functions. The basic requirement is that the interface agency as a whole should be able to organize itself in the way users' preferences and actual situation parameters call for. To achieve an ongoing adaptation by a negotiation technique, a simple representation of credits or a simple selection process of contractors is not suitable. To this end, general heuristics has to be defined which concerns the adjustment of credits and the assignment of agents. More concrete, the following requirements should be satisfied:

1. Credits represent the strengths of contractors to allow an intelligent assignment of the best contractor.
2. Current situation parameters constrain the adjustment of credits to model the influence of situational circumstances on users' preferences.
3. Contractors store different credits for different contractor-manager relationships to distinguish between managers.
4. On negative feedback, the direct activation of a contractor that has not caused the undesired solution is possible.
5. Time-varying preferences can be handled.
6. Assignment of dominant contractors from any number of contractors and by any number of managers can be realized.
7. To reduce communication overhead, tasks are allocated to a contractor directly if the contractor was successful in the preceding session.
8. Conflicts caused by the fact that one contractor is equally successful as other contractors can be resolved.

The first three aspects concern the contractor function; the last three aspects concern the manager function; aspects four and five pertain to both functions. Since agents of the entire multi-agent system can take on the role of a contractor as well as the role of a manager in different situations and at different time steps, each agent has to perform both functions. The execution of these functions is part of a communication and cooperation framework, so that internal functionalities can be realized independently, and no supervision by any kind of a globally informed agent is needed. In the following two sections, we describe in detail the heuristics to define both functions, the contractor and the manager function.

5 Agents as Contractors

To adapt to users' preferences, the contractor learning functions realize the adjustment of credits based on data extracted from negotiation messages and the manager learning functions realize the assignment of most promising contractors. In detail, each contractor function performs the following steps: extract and store relevant information from incoming negotiation messages, compute actual situation parameters, and use the results to adjust credits which are included in their bids.

5.1 Representation of Credits

The basic idea to decide which information is relevant for the adaptation task and, thus, has to be captured is motivated by the way tasks would normally be assigned in a company. Applied to our context, a contractor is considered successful when (1) a high number of tasks was sent to the contractor's sub-agency, (2) a high number of tasks was performed by that contractor, and (3) a high number of tasks was successfully performed by that contractor.

Realizing this idea demands that each contractor acquires knowledge at any negotiation step and represents the knowledge in some kind of a time-dependent vector of which it keeps track during the ongoing session. In fact, two kinds of vectors have to be defined. First, a data vector is used which represents acquired message data to determine the three above numbers. At any negotiation process t each contractor stores the following data vector[2]:

$$data(t) = (message\_id(t), sender(t), recipient(t), type(t), content(t))$$

where *type* refers to the negotiation technique used (task announcement or direct task allocation) and *content* represents the kind of feedback the user has given, i.e. positive or negative.

Second, a vector of credit values, in short, credit vector, is used which represents the agents' strengths at discrete negotiation steps including the three numbers mentioned above. At any interaction step t the credit vector of a contractor is defined as follows:

$$credit(t) = (conf(t), sit(t), suc\_task(t), perf\_task(t), task(t))$$

where *task* is the number of tasks sent to the contractor's sub-agency, *perf_task* is the number of tasks performed by that contractor, and *suc_task* is the number of tasks successfully performed by that contractor. *sit* represents the conformity of the current situation with the preference embodied by the contractor. *conf* results from combining the other vector components in a special way and represents the overall confidence the contractor has doing the task successfully.

In general, contractors cooperate with several managers. Since it is possible that a contractor has successfully performed tasks allocated by one manager but was unsuccessful performing tasks allocated by another manager, each contractor stores a credit vector for each manager and manipulates each vector depending on the current manager.

5.2 Computation of Situation Parameters

Naturally given users' preferences may depend on situational circumstances what can result in a variation on their preferences. For example, in our application scenario (cf. Section 7) the orientation of reference objects to the user's view is a relevant situation parameter. To handle this situational aspect, current situations have to be taken into consideration to realize user adaptation. In our approach, this is achieved by integrating the computation of situation parameters in the

---

[2] Precisely, a data vector $data_{sa}^c(t)$ is computed for each contractor c in any sub-agency sa involved. For the ease of reading, we have dropped indices c and sa in the formula. The same holds true for $credit(t)$, $sit(t)$, $task(t)$, etc.

adjustment of credits. That means that the current situation has to influence the adjustment of credits positively whenever this situation seems to be in conformity with the user's preference that the contractor is realizing.

To analyze the situations and integrate the results in the adaptation process, a set of independent features

$$f_1, f_2, \ldots, f_n$$

is used. Each feature has a value f_i^{opt} that is given when a situation confirms this feature optimally with respect to the preference the agent realizes and a current value $f_i(t)$ that determines the value of that feature at the negotiation step t. The kind of features and the number of features is defined locally by each agent and can be considered as part of its local knowledge. Features vary over sub-agencies but are equal within one sub-agency since each sub-agency models the same functionality type. Optimal feature values also vary within a sub-agency since each member of this sub-agency realizes another preference with other basic requirements of the modelled preference class.

On this basis, the current situation is then measured by computing the weighted sum of feature deviations:

$$\sum_{i=1}^{n} w_i \Delta f_i(t) = \sum_{i=1}^{n} w_i |f_i^{opt} - f_i(t)|$$

where the weight w_i determines the importance of the feature f_i as measurement for a situation. The result of this equation is a measurement for the deviation between the current situation and a situation where the preference is confirmed optimally. Since the component sit of the credit vector is a measurement for the conformity of the current situation with the preference embodied by a contractor, at any negotiation step t sit is computed as:

$$sit(t) = 1 - \sum_{i=1}^{n} w_i \Delta f_i(t)$$

The value of sit will be calculated whenever a contractor wants to generate a bid in response to a task announcement. Especially at the beginning of a session (when no or few information is available), the consideration of how well the current situation confirms a given preference is useful to make more promising predictions. with respect to adaptation efficiency. In the following section, we describe how sit is further used and integrated in the adjustment of credits.

5.3 Adjustment of Credits

So far, we described what kind of information is represented and how the situational influence on users' preferences is measured. For a given credit vector $credit(t) = (conf(t), sit(t), suc\_task(t), perf\_task(t), task(t))$, we now explain how the last three credit components and the confidence value are computed. The values of the last three vector components are computed on the basis of the information stored in the data vector and acquired at each interaction step t. Using this information, *task*, *perf_task*, and *suc_task* are updated by the following rules where c could be any contractor of a sub-agency sa. The value of *task* is incremented by 1 whenever a task is posted to a sub-agency sa or a task is directly allocated to a contractor c; otherwise it remains the same.

$$
task(t) := \begin{cases} 0, \text{if} & t = 0 \\ task(t-1) + 1, \text{if} & (type(t) = \text{task announcement} \wedge \\ & recipient(t) = sa) \vee \\ & (type(t) = \text{task allocation} \wedge \\ & recipient(t) = c) \\ task(t-1), \text{else} \end{cases}
$$

The value of *perf_task* is determined in a similar but a bit more restricted way such that it will be incremented whenever a contractor c has actually performed the corresponding task; otherwise it remains the same.

$$
perf\_task(t) := \begin{cases} 0, \text{if} & t = 0 \\ perf\_task(t-1) + 1, \text{if} & (type(t) = \text{task allocation} \wedge \\ & recipient(t) = c) \\ perf\_task(t-1), \text{else} \end{cases}
$$

The number of tasks successfully performed is incremented by 1 whenever a task was successfully performed by a contractor c, that is, the user feedback is positive; otherwise it remains the same.

$$
suc\_task(t) := \begin{cases} 0, \text{if} & t = 0 \\ suc\_task(t-1) + 1, \text{if} & (type(t-1) = \text{task allocation} \wedge \\ & recipient(t-1) = c \wedge \\ & content(t-1) = \text{positive}) \\ suc\_task(t-1), \text{else} \end{cases}
$$

Confidence. Confidence refers to the trust a contractor has doing the task successfully. Having computed the last three components and the *sit* component of the credit vector as described above, the value of *conf* can be updated by using these results in a combined way. The confidence a contractor has that it will execute the posted task successfully is influenced by three factors:

P: Performance with respect to the previous task announced to the contractor's sub-agency
D: Degree of the contractor's dominance in the preceding session
S: Degree of situational conformity

Speaking metaphorically, these three factors describe factors that influence and constrain the selection of users' preferences in the way that they reflect the user's satisfaction with the offered solution, the interaction history, and the situational dependency.

The performance of a contractor is the result of the user feedback and the contractor's involvement in the previous task. In more detail, the performance P at any negotiation step t is computed as:

$$P(t) \; := \; \begin{cases} 0, \text{ if } (task(t) = 0 \; \vee \; perf\_task(t) = 0) \; \vee \\ \quad (perf\_task(t) - perf\_task(t-1) = 0) \\ 1, \text{ if } (perf\_task(t) - perf\_task(t-1) = 1) \; \wedge \\ \quad (suc\_task(t) - suc\_task(t-1) = 1) \\ -1, \text{ if } (perf\_task(t) - perf\_task(t-1) = 1) \; \wedge \\ \quad (suc\_task(t) - suc\_task(t-1) = 0) \end{cases}$$

This means that the performance is high when a contractor has performed the previous task successfully; the performance is low when the solution which the contractor has offered was corrected by the user; the performance is neither high nor low when the contractor has either never performed before or when the previous task has not been performed by this contractor.

To determine the degree of the contractor's dominance, a test $dominant(t)$ is carried out which checks (1) if the number of successfully performed task in relation to the number of performed tasks (d_1) is greater than a threshold $\alpha(t)$ and (2) if the number of performed tasks in relation to the number of announced tasks (d_2) is greater than a threshold $\beta(t)$. The dominance degree D is based on the result of the $dominant$ predicate at any negotiation step t:

$$D(t) \; := \; \begin{cases} 0, \text{ if } (task(t) = 0 \; \vee \; perf\_task(t) = 0) \\ (d_1 + d_2)/2, \text{ if } dominant(t) \\ (d_1 + d_2)/2 - 1, \text{ if } \neg dominant(t) \end{cases}$$

The last influence factor S is based on the credit component sit in the way that a test $sit\_conform$ is performed which checks again if the value of sit is greater than a threshold $\gamma(t)$. The result of this test is used to compute the value of S at any negotiation step t in the following way:

$$S(t) \; := \; \begin{cases} sit(t), \text{ if } sit\_conform(t) \\ sit(t) - 1, \text{ if } \neg sit\_conform(t) \end{cases}$$

The thresholds α, β, γ are updated in correspondence to the contractors' performance P. When the performance is positive the thresholds are decreased; in the other case, the thresholds are increased. For example, the value of α is computed at any negotiation step t by the following rule:

$$\alpha(t) \; := \; \begin{cases} \alpha_{min}, \text{ if } \quad t = 0 \\ \alpha(t-1) - 1/n \, \triangle\alpha(t)P(t), \text{ else} \end{cases}$$

where $\triangle\alpha(t)$ determines the intervall remaining for manipulation. Thus, contractors need a lower dominance and a lower situational conformity to be considered as successful in future interactions when they have performed task positively in comparison to executing tasks negatively.

Resulting from the above computations, the confidence value at any negotiation step t is determined by:

$$conf(t) := \begin{cases} P(t), \text{if} & P(t) = -1 \\ (w_p P(t) + w_d D(t) + w_s S(t)), \text{else} \end{cases}$$

The rationale behind the definition above is as follows. A contractor which has a low performance as result of having received negative feedback gets the least minimal confidence value possible. In the other case, its confidence is given by computing the weighted sum of the three influence factors P, D, and S[3].

Combining these three factors allows to efficiently assess contractors just by considering one value. In a similiar way, [LMM94] use a trust value to select appropriate agents for collaboration. Nevertheless, the credit vector contains additional components that allow a more detailed discrimination between contractors in conflicting cases without solely relying on contractors' evaluations.

Defining credits in this way, the requirements described in Section 4 can be satisfied as far as they concern agents as contractors. (1) The credit vector, especially the *conf* value, represents a kind of strength and captures information that can be used for a more stable adaptation; (2) the current situation is integrated in the adjustment of credits; (3) different contractor-manager relationships are modelled; (4) the direct activation of another contractor on negative feedback as well as (5) the adaptation to time-varying preferences is prepared by the definition of *conf*.

6   Agents as Managers

Based on the contractor function, the manager function of each agent has to perform the following steps to achieve the overall adaptation to users' preferences: decompose and announce tasks, pool incoming bids corresponding to the task announcement, and assign the most promising contractor based on the results of the comparison of bids.

6.1   Announcement of Tasks

Each user's instruction requires the solution of a number of sub-tasks by the agent system as a unit. Therefore, a number of negotiation processes are carried out between agents within and across sub-agencies. To solve a sub-task, a manager usually gets other agents or sub-agencies involved which supply the manager with sub-results. Thus, the manager decomposes a sub-task in further sub-tasks and generates a task announcement for each task to be executed.

Besides general information including the sender, the send time, and the task description, a task announcement contains the user feedback of the current instruction. Negative user feedback is included explicitly whereas positive feedback

---

[3] In our current implementation, the weights each have the fixed value of 1/3.

is included implicitly. More precisely, negative feedback is indicated by appending a negative instruction in front of the task description which is left out in the case of positive feedback.

The question remaining is to which agents credit has to be assigned (commonly known as *credit-assignment problem*). In multi-agent learning, this problem decomposes into the problems of inter-agent and intra-agent credit-assignment [Wei96]. Concerning the inter-agent problem, the agent system has to decide what actions by what agent contributed to the performance change. In our approach, this decision can easily be made since users' preferences are considered as independent. This means that each user's instruction concerns one preference, i.e., a special sub-agency. Thus, the involved sub-agency can easily be determined and credited. Nevertheless, our approach can also handle preferences which depend on each other by assigning credit or blame to any sub-agency that contributed to the offered solution.

With regard to the intra-agent problem, an agent has to decide which facts led to the contributing action. In our case, the activation of a special agent of the contributing sub-agency relies on the credits included in bids. This means that this kind of problem is solved by adjusting internal credit values in correspondence to the preceding negotiation processes and the current feedback included in the task announcement. More generally, the inter-agent credit-assignment problem is solved by agents as managers whereas the intra-agent problem is solved by agents as contractors.

The execution of tasks by the way of announcing tasks requires a number of communication and cooperation processes. To reduce this interaction overhead, managers allocate tasks directly if contractors have executed tasks successfully over a period of contracts. Therefore, each manager acquires knowledge about contractors it is collaborating with and about their success or failure in the preceding session. This information is captured in a time-dependent history vector for each sub-agency[4]:

$$history(t) = (task(t), success(t), recipient(t), \delta(t))$$

where *task* represents the number of tasks allocated to a sub-agency, *success* stores whether the last task was performed positively or negatively, *recipient* refers to a specific contractor of the considered sub-agency, and δ corresponds to a time-varying threshold which is used for bid evaluation tests.

Using this knowledge, a manager can decide if a contractor has performed tasks successfully over a period of contracts and in the positive case allocate the next task to this contractor directly. A similiar idea was presented by [Dow95] where a learning contract-net algorithm is used to reduce communication overhead.

---

[4] For the ease of reading, the index *sa* indicating a special sub-agency is dropped in the formula; cf. footnote 1.

6.2 Pooling Bids

After having sent the task announcement, a manager waits for corresponding bids of the contractors of the collaborating sub-agency. Each incoming bid is preprocessed by

1. extracting credits included in the bid,
2. pre-computing factors which are relevant for comparison of bids, and
3. pooling the results in an internal credit data structure.

The factors mentioned by step two are results of tests concerning how successful the contractor that has sent the bid has been in the preceding session. Similiar to the way agents as contractors perform tests for computing their confidence, a manager performs tests that are used to determine the contractors' strength. For instance, it checks if the number of successfully performed tasks in relation to the number of performed tasks is greater than a threshold. To perform these computations, the threshold of the *history* vector is used. Before performing the tests, δ is updated by searching for the appropriate entry of the *history* vector and using the entry values in the following way:

$$\delta(t) := \begin{cases} \delta_{min}, \text{ if } t = 0 \\ \delta(t-1) - 1/n \left(\delta(t-1) - \delta_{min}\right), \text{ if } success(t) \\ \delta(t-1) + 1/n \left(\delta_{max} - \delta(t-1)\right), \text{ if } \neg success(t) \end{cases}$$

The pooling of bids is constrained by two disjunctive conditions that terminate the waiting process. The manager is waiting for incoming bids until a maximal time interval is reached or all members of the contracting sub-agency have sent their bids. Therefore, the manager determines how many contractors are active before sending the task announcement. As soon as one of these conditions are satisfied, the manager starts to compare bids or, if no bid was sent, stops the current negotiation process.

6.3 Comparison of Bids

To determine the contractor that has worked most successfully in the preceding session, a manager compares each bid with each other bid on the basis of the credit information extracted and pooled in the previous step. For efficiency, this procedure is not executed if just one bid was pooled. In more detail, the manager starts the comparison by considering the entire set of bids and, then, reduces the search space step by step using special criteria, described below, until precisely one contractor is left or all citeria have been matched. In the latter case, one contractor of the remaining set of contractors is chosen.

The criteria for assigning a contractor decompose in four main classes of rules. The first class of rules concerns the consideration and comparison of the confidence values, and includes the following three rules:

Rule 1:
Remove the bids where *conf* has the minimal value;
If the resulting set of bids is empty
      do stop;
else
      do *Rule 2*;

Rule 2:
Determine the maximal *conf* value;
If this value is greater than 0
      do *Rule 3*;
else   do *Rule 4*;

Rule 3:
Choose the bid where *conf* is maximal;
Stop;

Evaluating the confidence value *conf* first allows that another contractor can be activated when negative feedback occurs because bids with minimal confidence value will be removed first. If the maximal confidence value is not positive, the manager does not rely on the contractors' self-evaluation and, thus, proceeds with further evaluations.

The second criterion considers the evaluation of the situational conformity by performing rule four:

Rule 4:
Choose the bids where *sit* is greater than δ;
If the resulting set contains exactly one bid
      do stop;
else
      do *Rule 5*;

Satisfying that the *sit* value is greater than a threshold, a manager allocates a task to a contractor even if its other credit values, especially its confidence value, are low. By this, time-varying preferences based on situational circumstances can be modelled.

The third class of rules similiarly aims at modelling time-varying preferences but concentrates on the degree of dominance which contractors have reached.

Rule 5:
Choose the bids where the relation between *suc_task* and *perf_task* (q_1) and the relation between *perf_task* and *task* (q_2) is greater than δ;
If the resulting set contains exactly one bid
      do stop;
else if the resulting set contains more than one bid
      do *Rule 6*; **else**
      do *Rule 7*;

The rationale behind *Rule 5* is that contractors should be preferred if they have worked better than average in the preceding session. A manager discriminates between the situations where none of the contractors or more than one contractor satisfies this condition (and/or the condition of *Rule4*) since it makes a difference whether to choose between poor contractors or between better ones.

The last criterion, finally, contains rules to distinguish between contractors in order to determine most promising ones. Thus, these rules can be regarded as conflict resolution rules.

Rule 6:
Determine the bids where $\sum (conf + sit + q_1 + q_2)$ is maximal;
If the maximal value is ambiguous within an ε-intervall
      do choose the bids where *sit* is maximal;
else
      do stop;

Rule 7:
Choose the bids where the difference between *perf_task* and *suc_task* is minimal;
If the resulting set contains more than one bid
      do *Rule 6*;

The first of these rules distinguishes between more promising contractors by taking almost all credit values into account and, in conflicting cases, preferring the one where the situational conformity is maximal. By the latter rule, the manager first chooses that contractor which has caused a minimal number of negative feedbacks during the preceding session (to determine the best one of the poor ones). Subsequently, the manager performs *Rule 6*, described above, to handle conflicting cases.

Defining rules in this way, the direct activation of another contractor on negative feedback, the handling of time-varying preferences, the assignment of successful contractors from any number of contractors, and the discrimination of contrators in conflicting cases can be realized. Moreover, tasks are allocated to contractors directly to reduce communication overhead. Thus, the requirements stated in Section 4 are satisfied completely.

7 Example Application

The adaptation method described in this paper was implemented in a multiagent interface system for interaction with a 3D-virtual environment, carried out in the VIENA project [WC95]. The user can instruct the system by way of verbal and gestural input. These qualitative instructions are translated by the interface agency to internal commands which can be interpreted by the graphical system. As an example application, a virtual office room can be manipulated by the user. To enhance interaction comfort, we have realized an anthropomorphic agent, named Hamilton, that is visualized in the scene (Figure 3) and can respond to and carry out users' instructions [WLJJLF].

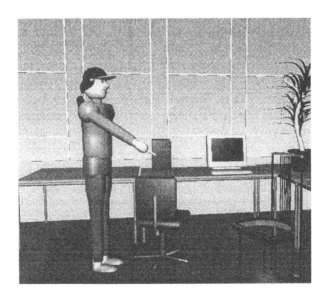

Fig. 3. Snapshot of a VIENA example scene

7.1 VIENA Interface Agency

To reconstruct the overall user's instruction, a number of different tasks have to be solved which are distributed within the multiagent interface system. The entire interface agency consists of a set of communicating and cooperating sub-agencies, each of them realizing a different functionality (cf. Figure 4). For instance, a space agency determines spatial transformations, a color agency changes the appearance of scene objects, a hamilton agency determines actions

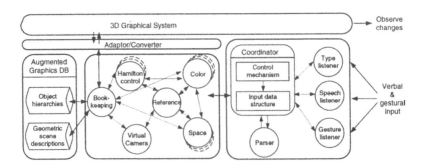

Fig. 4. Architecture of the VIENA multiagent interface system

of the anthropomorphic figure. Agents communicate and cooperate by using a variation of the contract-net negotiation protocol in which each agent can take on the role of a contractor as well as a manager [LWC95].

Since the user interacts with the system by way of qualitative verbal and gestural instructions (which are often situated), their precise meanings can usually not be resolved unambiguously. Rather different solutions are possible. The practical experience with the VIENA system has shown that significant variations of users' preferences exist with respect to possible solutions. Thus, the VIENA agency proves to be a reasonable testbed for integrating adaptation facilities. We have built agents of the same type but with slightly different internal functionalities – corresponding to different users' preferences – and joined them together in a sub-agency. To this end, the requirements for using the contract-net learning approach (cp. Section 3) are satisfied.

7.2   Contract-net Learning in VIENA

The adaptation method described in this paper has been implemented and tested for a variety of examples, primarily for the case of users' preferences for different spatial reference frames. The semantics of spatial instructions may depend on different perspectives [Ret88], e.g., they may be interpreted from the user's point of view (deictic perspective) or, alternatively, from the point of view of an object which has a prominent front (intrinsic perspective). Experiments with the VIENA system have shown that, due to individual differences among users,

Fig. 5. Possible solutions of the instruction "Hamilton, go left.": Hamilton can move to the left from the hamilton-intrinsic (H) or the user-deictic (U) perspective.

one spatial reference frame may be preferred over the other one [JW96].

As an example, consider the situation in Figure 5. Two possible solutions can be offered by the system when the user has instructed Hamilton to go left. In addition, users' preferences for spatial reference frames may depend on the orientation of the object given in the actual situation. On the basis of the hamilton-intrinsic reference frame, Hamilton would move in the direction indicated by 'H'; on the basis of the user-deictic reference frame, Hamilton would move in the direction indicated by 'U'. Which reference frame is preferred depends on the preference of the current user in the given situation. Therefore, we have implemented two instances of the Hamilton sub-agency and equipped them with the contract-net learning framework. Whenever the user gives an instruction concerning the anthropomorphic figure, both contractors compute situation parameters, adjust their credits, and generate a bid including credits. Situation features concern the orientation of Hamilton and the distance between Hamilton and the user (camera). Any other agent of the interface agency, e.g. the coordinator agent, can act as a manager.

Similarly, we have implemented two space contractors which compute spatial transformations on the basis of the user-deictic reference frame, or on the basis of the object-intrinsic reference frame, respectively. In addition to the two situation features mentioned above, the space agency uses three further features to analyze the situations: intrinsic character of the reference object and of the possibly underlying object as well as the orientation of the underlying object.

Finally, we have tested the adaptation method for the case of users' preferences for different color sensation by implementing two color agents that offer more drastic or smoother color transformations. Situation features could pertain to the lightings of the scene but are, so far, not integrated in the system.

First experiments have shown that the approach described above can realize adaptation to users' preferences effectively and satisfactorily with respect to the requirements stated in Section 4. A fuller evaluation, to be carried out on the basis of artificial users, is one of our very next goals.

8 Discussion

This paper presented an multi-agent learning approach to user adaptation realized by an interface agency. The interface agency consists of several sub-agencies which represent different pre-established preference classes. Each sub-agency consists of several agents corresponding to the possible preferences. Agents use a contract-net based negotiation process where each agent can take on the role of a contractor as well of a manager. As contractors, agents acquire knowledge about the user, the preceding session, and situation parameters, which is captured in internal credit vectors. As managers, agents acquire knowledge about collaborating contractors and about their success or failure in the preceding session. By learning from indirect user feedback in the ongoing session, contractors and managers compete with each other to meet the users' preferences.

Besides extending and using the contract-net negotiation as an intuitive mechanism to achieve user adaptation, one of our main results is a general framework for adaptation to users' preferences which can easily be integrated in a contract-net based multi-agent system. By this, internal functionalities can be realized independently, and no supervision by any kind of a globally informed agent is needed. Furthermore, the system's knowledge of the user is expressed in the activation of certain agents of the entire interface agency. In this way, user adaptation is achieved by the use of implicit, distributed user models, without accumulating explicit user models.

In the future, our main working topics will concentrate on the evaluation of the learning strategy. In order to carry out replicable experiments, we will define artificial users and compare the results offered by the interface agency with predefined expectations of the artificial users. This evaluation will include the detailed evaluation of the contractor and manager function and its optimization.

References

[Chi90] Chin, D.N. Intelligent Interfaces as Agents. In Sullivan, J.W. & Tyler, S.W. (eds.): *Intelligent User Interfaces* (pp. 177-206). New York: ACM Press, 1990.

[DS83]   Davis, R., Smith, G. Negotiation as a Metaphor for Distributed Problem Solving. In Bond, A.H. and Gasser, L. (eds.): *Readings in Distributed Artificial Intelligence* (pp. 333–356). Morgan Kaufmann, 1983.

[Dow95] Dowell, M.L. Learning in Multiagent Systems. Ph.D. Thesis at the Department of Electrical and Computer Engineering, University of South Carolina, 1995.

[JW96]   Jörding, T., Wachsmuth, I. An Anthropomorphic Agent for the Use of Spatial Language. To be published in Olivier, P. & Maass, W. (eds.): *Vision and Language*, Springer, 1997.

[Kay90] Kay, A. User Interface: A Personal View. In Laurel, B. (ed.): *The art of human-computer interface design* (pp. 191-208). Reading: Addison-Wesley, 1990.

[LMM94] Lashkari, Y., Metral, M., Maes, P. Collaborative Interface Agents. In *Proceedings of the National Conference on Artificial Intelligence*. Cambridge (MA): The MIT Press, 1994.

[Lau90] Laurel, B. Interface agents: Metaphors with character. In Laurel, B. (Ed.): *The art of human-computer interface design* (pp. 355-365). Reading: Addison-Wesley, 1990.

[LW96]   Lenzmann, B., Wachsmuth, I. A User-Adaptive Interface Agency for Interaction with a Virtual Environment. In Weiss, G. & Sen, S. (eds.): *Adaption and Learning in Multi-Agent Systems* (pp. 140-151). Berlin: Springer, 1996.

[LWC95] Lenzmann, B., Wachsmuth, I., Cao, Y. *An Intelligent Interface for a Virtual Environment*. KI-NRW (Applications of Artificial Intelligence in North-Rine Westphalia) Report 95-01, 1995.

[Mae94] Maes, P. Agents that Reduce Work and Information Overload. *Communications of the ACM 37*(7), 1994, 31-40.

[MK93]   Maes,P., Kozierok, R. Learning interface agents. In *Proceedings of the Eleventh National Conference on Artificial Intelligence* (pp. 459-465). AAAI Press/The MIT Press, 1993.

[McT93] McTear, M.F. User modelling for adaptive computer systems: a survey of recent developments. *Artificial Intelligence Review* 7, 1993, 157-184.

[MCFMZ94] Mitchell, T., Caruana, R., Freitag, D., McDermott, J., Zabowski, D. Experiences with a learning personal assistent. *Communications of the ACM* 37(7), 1994, 80-91.

[Nor94] Norman, D.A. How Might People Interact with Agents. *Communications of the ACM* 37(7), 1994, 68-71.

[OHA96] Ohko, T., Hiraki, K., Anzai, Y. Learning to Reduce Communication Cost on Task Negotiation among Multiple Autonomous Mobile Robots. In Weiss, G. & Sen, S. (eds.): *Adaption and Learning in Multi-Agent Systems* (pp. 177-190). Berlin: Springer, 1996.

[Ret88] Retz-Schmidt, G. Various views on spatial prepositions. *AI magazine* 9(2), 1988, 95-105.

[Sel94] Selker, T. Coach: A Teaching Agent that Learns. *Communications of the ACM* 37(7), 1994, 92-99.

[WC95] Wachsmuth, I., Cao, Y. Interactive Graphics Design with Situated Agents. In W. Strasser & F. Wahl (eds.): *Graphics and Robotics* (pp. 73-85). Berlin: Springer, 1995.

[WLJJLF] Wachsmuth, I., Lenzmann, B., Jörding, T., Jung, B., Latoschik, M., Fröhlich, M. A Virtual Interface Agent und its Agency. Poster contribution to the *First International Conference on Autonomous Agents, Agents-97*.

[Wei96] Weiß, G. Adaptation and Learning in Multi-Agent Systems: Some Remarks and a Bibliography. In Weiß, G. & Sen, S. (eds.): *Adaption and Learning in Multi-Agent Systems*. Berlin: Springer, 1996.

The Communication of Inductive Inferences

Winton Davies and Peter Edwards

Department of Computing Science,
King's College,
University of Aberdeen,
Aberdeen,
Scotland, UK, AB24 3UE

{wdavies, pedwards}@csd.abdn.ac.uk

Abstract We propose a new approach to communication between agents that perform inductive inference. Consider a community of agents where each agent has a limited view of the overall world. When an agent in this community induces a hypothesis about the world, it necessarily reflects that agent's partial view of the world. If an agent communicates a hypothesis to another agent, and that hypothesis is in conflict with the receiving agent's view of the world, then the receiving agent has to modify or discard the hypothesis.

Previous systems have used voting methods or theory refinement techniques to integrate these partial hypotheses. However, these mechanisms risk destroying parts of the hypothesis that are correct. Our proposal is that an agent should communicate the bounds of an induced hypothesis, along with the hypothesis itself. These bounds allow the hypotheses to be judged in the context from which they were formed.

This paper examines using version space boundary sets to represent these bounds. Version space boundary sets may be manipulated using set operations. These operations can be used to evaluate and integrate multiple partial hypotheses. We describe a simple implementation of this approach, and draw some conclusions on its practicality. Finally, we describe a tentative set of KQML operators for communicating hypotheses and their bounds.

1. Introduction

In this paper we address the question: *"How should agents communicate inductive inferences?"* This question is important because of a growing interest in systems that distribute a learning task amongst a community of agents (Weiss and Sen, 1995). Such systems are often referred to as multi-agent learning systems (Brazdil, 1991). The question is also relevant to the development of the Knowledge Query and Manipulation Language (Finin, 1993). KQML currently does not define standards for the communication of inductive inferences. Inductive inference differs from deductive inference in that it is logically unsound. This means that a logical sentence derived using inductive inference with respect to partial knowledge of a world may be inconsistent with the world as a whole. This is the central problem when communicating an induced hypothesis. In a multi-agent learning system this requires agents to evaluate or refine hypotheses proposed by other agents. The major problem with this is that the refinement process may change the rule so that whilst it is correct for the refining agent, it may no longer be correct with respect to the original agent's view of the world.

We were initially motivated by a desire to extend KQML so that it could be used for communication within multi-agent learning systems. All previous systems have used a variety of messages to facilitate communication between learning agents. We had intended to rationalise these into a core set of KQML message primitives (or performatives). These would have been performatives such as "learn", "refine", and "test". However, the following question arose: "How could inductive inferences be communicated, whilst simultaneously maintaining a logically sound distributed knowledge-base?" This led us to propose a solution based on the communication of the bounds to the inductive inference."

Version spaces (Mitchell, 1978) seem the natural candidate to describe these bounds, as they represent all possible hypotheses that are consistent with an agents' view of the world. In Section 2.2 we provide an overview of Mitchell's work and review more recent extensions to it. Briefly, Mitchell showed that for a given supervised inductive learning problem, that the space of consistent hypotheses may be represented by two sets, one containing the most general hypotheses and the other containing the most specific hypotheses. These are jointly called the version space boundary sets.

This paper examines the feasibility of using version space boundary sets to represent the bounds of an inductive hypothesis. Thus an agent that receives an induced hypothesis and the bounds now knows the space of hypotheses that was consistent for the sending agent, as well as the particular hypothesis that was selected. The receiving agent is now able to modify the specific hypothesis if it is inconsistent with its own world view, but can constrain the modifications it makes to be consistent with the set of hypotheses that was valid for the originating agent. We shall now give a simple example of the problem and the solution.

1.1 A Simple Example

In order to demonstrate the problem of agents communicating inductive hypotheses, we consider the simple goal of learning a propositional binary relation. Even this trivial task (there being just 16 possible hypotheses) demonstrates the essential elements involved. We describe the learning task, how it is represented as a version space and then show how it is affected when distributed between two agents. In this example we are only considering distribution of the training examples. This is referred to as horizontal distribution. However, other aspects of the learning problem can be distributed. These include vertical distribution of examples (agents each have different facts about all the examples), as well as the distribution of language biases and selection biases.

1.2.1 A Simple Supervised Learning Problem

The task is to learn an intensional definition of a relation $f(X)$ in terms of two unary predicates (e.g. $p(X)$ and $q(X)$) and the full range of Boolean operators (i.e. $\neg, \wedge, \vee$). We are given positive and negative extensional examples of $f(X)$, and the Boolean value of $p(X)$ and $q(X)$. Thus there are 4 distinct possibilities for any particular example (corresponding exactly to the traditional propositional truth table shown in Table 1).

X	p(X)	q(X)	f(X)
0	F	F	?
1	F	T	?
2	T	F	?
3	T	T	?

Table 1 A truth table for features p(X) and q(X)

There are 2^4 (i.e. 16) possible hypotheses for the definition of $f(X)$. However, once we have seen a positive or negative example for each one of the four truth table rows, then the relation will be known. However, given less than the four distinct examples, an inductive leap must be made. This simple example shows the inputs to an inductive learning algorithm:

1. The conceptual bias is simply $p(X)$ and $q(X)$.

2. The language bias is all possible logical combinations of the conceptual bias. Together 1 and 2 constitute the concept description language (CDL).

3. The selection bias is the preference the algorithm has in selecting one of the remaining set of possible hypotheses which are consistent with the training examples. Typically Occam's Razor is used, which selects the shortest syntactic hypothesis (e.g. given a choice of $p(X) \Rightarrow f(X)$ or $p(X) \wedge q(X) \Rightarrow f(X)$ it would chose the former).

4. The training examples are of the form $f(object-1) \in \{T, F\}$, $p(object-1) \in \{T, F\}$, $q(object-1) \in \{T, F\}$.

1.2.2 Representing the Problem as a Version Space

We can now show the version space corresponding to this example (see Fig 1). Each node of the version space represents a hypothesis. A hypothesis can be represented by the extensional definition of the relation $f(X)$, i.e. as a set that contains the distinct values of X such that $f(X)$ is true. The hypothesis is equivalently represented by an intensional description in terms of $p(X)$ and $q(X)$. In Fig 1, each node contains both forms. Each node contains a description of the form $\{2,3\} = P$, where $\{2,3\}$ represents the extensional form, and P represents the intensional form of the hypothesis.

A hypothesis *h1* is more general (or more specific) than a hypothesis *h2*, iff *h1* ⊂ *h2* (or *h1* ⊃ *h2*). The lattice connecting the nodes is a partial order of generality between hypotheses. This partial order is most clearly seen in the extensional form. With this particular concept description language (CDL), there are only 4 distinct objects in the universe, which are denoted by n ∈ {0..3}. Thus the hypothesis ¬*p(X)* ∧ ¬*q(X)*, covers a single object ({0}), which is more general than bottom node (described as *F*, for all false), covering no objects ({ }), and directly more specific than the hypotheses ¬*p(X)*, ¬*q(X)*, and (¬*p(X)* ∧¬*q(X)*) ∨ (*p(X)* ∧ *q(X)*), which cover {0, 1}, {0, 2}, and {0, 3} respectively.

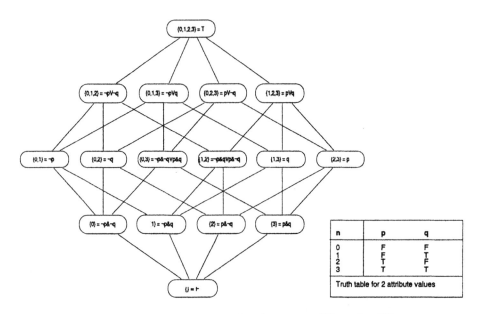

Fig 1: A version space for features p(X) and q(X).

A version space allows us to express the set of valid hypotheses in terms of boundary sets consisting of the most specific generalisations (G) and most general specialisations (S). Normally G and S will be described intensionally. The ability to represent a set of hypotheses by the boundary sets is the essence of the power of the version space approach.

Let us now look at how the version space changes with training examples (see Fig 2). Initially, G is set to *T* (standing for "all true", the top node), and S is set to *F* (the bottom node). Now these describe the initial version space, because all 16 hypotheses are valid with respect to no training examples. Imagine now, that we have a negative training example f(object-1) where p(object-1) and q(object-1) are both true. This changes G from *T* to ¬*p(X)*∨¬*q(X)*, leaving only 8 possible hypotheses.

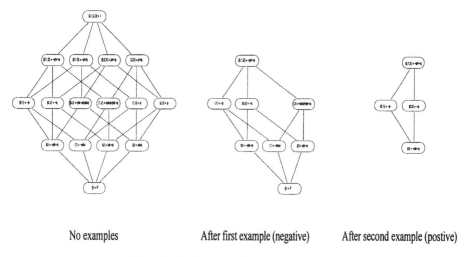

| No examples | After first example (negative) | After second example (postive) |

Fig 2: Updating the version space.

Now if we have a positive example where $p(X)$ and $q(X)$ are false, then the S set is changed to $\neg p(X) \wedge \neg q(X)$. There are now only 4 possible hypotheses left (G, S, $\neg p(X)$, and $\neg q(X)$). Note that if G predicts an unclassified training example to be false, then it is, but if it classifies it as positive, then it may be wrong. S is the inverse, and until we have G = S, then we must guess if an unclassified example falls into the space between G and S. This basis for this guess is the selection bias. If Occam's razor is used, then we would select the shortest one, in which case is a randomly choice between $\neg p(X)$ or $\neg q(X)$.

Traditionally, a supervised concept formation learning algorithm such as ID3 (Quinlan, 1986), FOIL (Quinlan 1990), or FOCL (Pazzani, 1992), directly calculates a hypothesis, without considering G & S. However, by definition, even a directly calculated hypothesis has a version space. The basis of proposed approach is, in fact, that agents make this version space explicit.

1.2.3 Distributing the Learning Problem

Let us now turn our attention away from the version space representation of the hypothesis space, and examine distributed learning in the context of this simple example. Imagine two learning agents, each of which has two distinct sets of training examples. If all these examples were given to one agent, then the relation $f(X)$ would be fully defined, and the intensional definition would be sound. However, with only 2 examples each, the best each agent can do is to generate the set of 4 possible hypotheses each, and chose one according to its selection bias. For example, let us consider a situation in which the agents have the training examples shown in Table 2 and Table 3 below.

X	p(X)	q(X)	f(X)
2	T	F	F
3	T	T	T

Table 2: Examples for Agent 1.

X	p(X)	q(X)	f(X)
0	F	F	T
1	F	T	F

Table 3: Examples for Agent 2.

This example was generated using the FOCL program (Pazzani, 1991). For agent 1, the actual version space is $G_1 = \neg p(X) \vee q(X)$ and $S_1 = p(X) \wedge q(X)$. The selected hypothesis using FOCL, F_1, is $q(X)$. For agent 2, the actual version space is $G_2 = p(X) \vee \neg q(X)$ and $S_2 = \neg p(X) \wedge \neg q(X)$. The selected hypothesis, F_2 is $\neg q(X)$. So, what is the problem? Quite simply, each agent generates a hypothesis that is inconsistent with respect to all the examples. The correct (and sound) inference for the undistributed examples should be $(\neg p(X) \wedge \neg q(X)) \vee (p(X) \wedge q(X))$. If either hypothesis ($F_1$ or F_2) was tested against the other agent's two examples, it classify both examples incorrectly. So what is the solution ?

1. The simplest solution would be to send all the training examples to one agent. However, we are only dealing with distinct examples, but there could be a large number of duplicate examples. For example, let X be US Social Security ID's, and p(X) represent male or female, and q(X) represent employed or not. The hypothesis we are learning from examples may be female unemployed and male employed, and yet there could be approximately 2.2×10^8 training examples (the US population) , of which there are only 4 possible distinct examples given this conceptual bias.

2. The second solution is for one agent to send the induced hypothesis to the other agent, and then for this agent to perform theory revision on it, with respect to its own examples. We have attempted this using the theory revision mechanism of FOCL (Pazzani, 1991). Agent 1 sends its induced hypothesis $(q(X) \Rightarrow f(X))$ to agent 2, which uses it as background knowledge. Even given this information, FOCL returns $\neg q(X) \Rightarrow f(X)$ as the hypothesis for agent 2. What went wrong? There is a fundamental problem facing theory revision: the theory reviser cannot know what version space the background knowledge was selected from. The FOCL revision algorithm has to make an assumption as to the version space the rule is drawn from. A natural approach is to assume the theory is over specific and/or over general. In this case, we can see that the hypotheses the theory reviser can consider are $\{F, \neg p, \neg q, \neg p \vee q, p \vee q, T\}$, (see Fig 3). None of which fall within the version space of the second agent's examples.

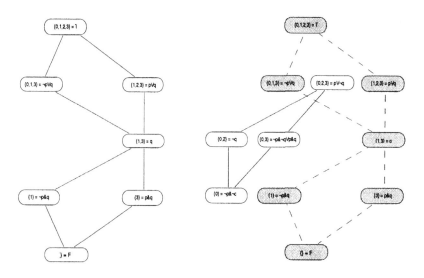

Version space derivable from agent 1 rule using theory revision.

Version space derivable from agent 2 training examples. Note no intersection.

Fig 3: Theory Revision and Version Spaces.

3. The third solution forms the basis of our approach. Agent 1 must send the version space that bounds its chosen hypothesis, as well as the hypothesis itself. Now the second agent may search this for a hypothesis consistent with its example (perhaps using the induced hypothesis as a starting point), or it may use version space intersection [Hirsh 89] to find the solution.

1.2.4 Intersecting Version Spaces

Let us finally demonstrating how the version space of agent 1 can be intersected with that of agent 2. A version space is a set, and thus can be intersected. There is a simple algorithm to do this if the CDL is unrestricted logically. This is simply to conjoin the G sets, and disjoin the S sets. Fig 4 shows an example of the intersection of the two version spaces from the previous example.

The first two version spaces show the set of possible hypothesis for each agent, and the one preferred by FOCL is highlighted. The final version space is the result of intersecting the two sets, which in this case is a singleton, containing the only possible hypothesis, $(\neg p(X) \land \neg q(X)) \lor (p(X) \land q(X))$. If the intersection is not a single hypothesis, then one again could calculate the preferred hypothesis, according to a particular selection bias.

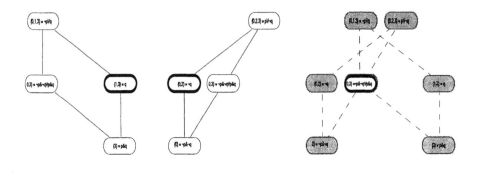

Version space based on agent 1's examples   Version space based on agent 2's examples   The Intersected Version Spaces

Fig 4: Intersecting Version Spaces

For completeness' sake, the symbolic form of the intersection is as follows:

$$
\begin{aligned}
G_{1 \wedge 2} &= (\neg p(X) \vee q(X)) \wedge (p(X) \vee \neg q(X)) \\
&= (\neg p(X) \wedge p(X)) \vee (p(X) \wedge q(x)) \vee (\neg p(X) \wedge \neg q(X)) \vee (q(X) \wedge \neg q(X)) \\
&= (p(X) \wedge q(X)) \vee (\neg p(X) \wedge \neg q(X)) \\
S_{1 \vee 2} &= (p(X) \wedge q(X)) \vee (\neg p(X) \wedge \neg q(X))
\end{aligned}
$$

As can be seen, G and S are identical, and so we detect the version space is a singleton.

1.3 Structure of the Remainder of the Paper

In Section 2 we describe some previous multi-agent learning systems and provide an overview of version space theory. We also review the Inferential Theory of Learning which we plan to use as the basis of our communication language (Michalski, 1993).

In Section 3 we describe our initial implementation, and discuss the practicality of the approach. We also discuss a number of features of boundary set representations which may prove useful.

Section 4 attempts to outline the KQML performatives to be used in communicating inductive inferences.

Section 5 lays out the future direction of this work. In particular we discuss how to increase the efficiency of our approach, how to handle integrating different learning biases, and how to deal with limited relational (i.e. first order) learning tasks.

2. Related Work

2.1. Multi-Agent Learning Systems

Brazdil et al. (1991) discuss a number of early multi-agent learning systems and attempt a classification of the general types. They defined a multi-agent learning system as one in which two or more agents communicate during learning. Their simplest class of system is when distributed data-gathering agents communicate with a single learning agent. Given that this trivially requires an algorithm to modify a concept with respect to new examples, then this class can be thought of as abstractly including all incremental learning algorithms, e.g. ID5 (Utgoff, 1989) and theory refinement systems, e.g. EITHER (Mooney & Ourston, 1991). The second class of systems integrate individual agent's hypotheses into a single theory. One method is to simply order the hypotheses (Gams, 1989). Another method is for the agents to vote for the best hypothesis (Brazdil & Torgo, 1990). The final class of system discussed by Brazdil et al. are hybrids, in which agents both refine and vote on hypotheses during the inductive process; see for example (Sian, 1991). It should also be noted that agents may communicate knowledge induced by one agent which is then used as background knowledge for another learning goal; see, for example (Davies, 1993).

Since this initial attempt to survey activity in the area, a number of additional systems and methods have been developed: Provost & Hennessey, 1995; Silver et al., 1991; Svatek, 1994; Chan & Stolfo, 1995. There have also been a number of workshops, as well as a comprehensive survey of learning in Distributed Artificial Intelligence (Weiss and Sen, 1995) Researchers in the field of computational learning theory have also analysed aspects of the area of multiple learners. These include Kearns and Seung (1995) who model inductive inference using an "oracle". Normally, each call to the oracle returns a single classified example. They show that to model a system with multiple learners, each call to the oracle should return a summary of the examples seen by a single agent. Cesa-Bianchi et al. (1995) also address a similar problem and present algorithms for integrating the results of "experts" (single agents that make a series of predictions over time). Finally, Jain & Sharma, (1993) address the issue of team learning, in which just one agent of a team must learn the correct concept.

2.2. Version Spaces

Mitchell (1978) first introduced the version space model of inductive inference, by demonstrating how inductive inference could be seen as a search for and through a set of hypotheses consistent with the available knowledge. This set of candidate concept definitions is called a *version space*. The basic components used to construct a version space are: positive and negative examples of the concept and representational biases (also called the Concept Description Language or CDL).

The first representational bias is the set of terms that can be used to describe the concept (often called the conceptual bias). The second representational bias is a restriction on the logical combination of these terms (often called the logical language). The examples together with the representational biases uniquely determine the version space of an inductive inference. The examples are used to eliminate all inconsistent hypotheses. The representational biases constrain the available hypotheses. The conceptual bias determines the maximum size of the version space. The logical language can decrease the size of this space, by disallowing certain hypotheses. It should be noted that background knowledge can be used deductively at any stage in this process: whether it is to determine the classification of an example or to deduce values of the terms used in the concept description language.

Mitchell demonstrated that a partial ordering exists over the hypotheses, based on the subset relation between the extensional definitions of hypotheses. This partial ordering defines a lattice which allows the version space to be represented using two *boundary* sets (G and S), which contain the most general and most specific hypotheses. As training examples are presented, the boundary sets are updated using the Candidate Elimination Algorithm (Mitchell 1978).

To complete the inductive inference a selection bias is applied to the version space. This imposes a preference order over the candidate hypotheses. An example of a selection bias is Occam's Razor, where hypotheses with shorter descriptions are preferred. However, it should be noted that all the hypotheses in the version space are valid concept definitions for the available examples. This is why maintaining version spaces keeps an inductive inference logically sound. If we select one hypothesis (using the selection bias), then we are uncertain if that concept definition is sound. However, if we maintain the set of all consistent concept definitions, it can be stated that exactly one of these definitions is sound. As more examples and/or representational biases are added, then the version space can converge to a singleton hypothesis, which is the logically sound definition of the concept. However the agent should never communicate a selected candidate hypothesis as if it had been soundly deduced until this point is reached.

It is also possible to obtain the same convergence effect by intersecting version spaces (Hirsh, 89). For a given learning goal, the examples and representational biases can be partitioned between several agents. Each agent then learns a version space consistent with the available knowledge. The resulting version spaces can then be intersected to form a new version space. This version space is identical to the one that would have been generated by a single agent with all the examples and representational biases available to it.

Hirsh extended the original version space model in several ways. Firstly, he defines the mathematical conditions (*convexity* and *definiteness*) that are necessary to represent the hypothesis space using boundary sets. Convexity occurs if a subset of a partially ordered set does not contain any "holes" with respect to the partial order. Definiteness occurs if a set has a finite number of elements at the upper and lower limits of the partial order. These properties taken together allow a hypothesis space to be represented using boundary sets (and thus the space is a version space). Hirsh proves that all finite Concept Description Languages are representable as version spaces, and we will therefore restrict our discussion to these languages.

Secondly, Hirsh proves that two version spaces may be intersected to form a new version space that is consistent with the two original version spaces. He defines an algorithm to perform the intersection. Hirsh suggests that this algorithm be used to incrementally learn a version space, by creating version space from single examples, and then intersecting them together. He also demonstrates how background knowledge can be incorporated, by generating a version space consistent with the knowledge, which can then be intersected with the version space being learned from the examples. This allows version spaces to be used for explanation-based learning (Mitchell et.al. 1986), as well as for concept learning. It is important to note here that any two version spaces can be intersected. By proving this property, Hirsh allows us to distribute the task of inductive inference, as described earlier.

Smith (1995) adopts a similar approach to Hirsh, but describes the integration process as a constraint optimisation problem. His aim is to simulate any inductive algorithm using a set of constraints and preferences. We will not examine his work in detail here, except to note that it addresses the issue of formally specifying the inductive biases. He shows how biases may be described using regular grammars (or finite state automata).

2.3. The Inferential Theory of Learning

Michalski's (1993) Inferential Theory of Learning describes a unified theory of learning. Simply put, the theory states that there are two types of inference: deductive and inductive, which give rise to other forms of reasoning such as analogy and abduction. These occur when measures of similarity are introduced. Michalski argues that a learning problem can be regarded as the application of inference-based operators to knowledge. In this model, there are two basic sets of knowledge: the reference set and the descriptor set. The reference set defines the input to the inference process (i.e. the training examples), whilst the descriptor set defines the output model (which we refer to here as the concept description language).

Operators may manipulate the reference set or descriptor set. Eight pairs of operators are defined which transform the knowledge in the system, and three pairs of operators which manipulate the knowledge (but do not transform it). The transformation operators include the following pairs which relate to inductive inferences:

- *Generalise & Specialise* change the number of objects covered by a concept definition;

- *Concretion & Abstraction* changes the number of hypotheses by varying the descriptor set;

- *Associate* and *Disassociate* increase or decrease the size of a reference set;

- *Characterise* and *Discriminate* define the target concept in terms of the reference set(s);

- *Select* and *Generate* either select a candidate hypothesis, or generate a set of hypotheses.

The three remaining operators relate to clustering, analogy, similarity based learning, and ad-hoc modifications of existing knowledge.

Michalski's central argument is that a learning agent can be designed which applies any of these operators to knowledge to achieve a given learning goal. We agree with his thesis, but have modified it to fit the version space model in a distributed environment. Thus the ITL combined with version spaces forms the basis of our proposed KQML operators in Section 4.

3 An Initial Implementation of Distributed Learning using Version Space Boundary Sets

This section describes a prototype learning agent that uses version space boundary sets to bound induced hypotheses. It both creates the boundary sets for communication, and uses received boundary sets to bound the refinement of hypotheses. Its basic learning algorithm is an attribute value version of FOIL (Quinlan, 1990). That is, it uses training examples in the form *attribute(example-id, value)*, and generates horn clauses that describe the given concept. We use the EPILOG theorem prover (Genesereth, 1995) to provide deductive inference capabilities.

Our intended model of distributed learning is as follows:

- Agents generate the hypothesis (F) that best fits their view of the training examples and boundary sets G & S that describe the version space of the training examples (and of course the CDL).

- One of the agents then receives all the other hypotheses and bounds and then performs the following steps:

  1. Conjoin all the G bounds and S bounds to produce G' and S'

  2. Check to see if any of the hypotheses obey the following relationship:
     $$S' \rightarrow F \rightarrow G'$$

  3. If one does, then this is the best hypothesis (F').

  4. Else use G' and S' to generate F'.

- The triple G', S', F' now represents the best hypothesis for the community as a whole.

Step 2 follows from the fact that the implication relational operator is the inverse of subsumption, and that the relationship defines membership of the version space. Step 3 follows from the definition of selection bias as a partial ordering of the hypotheses. Step 4 makes use of a heuristic, that the shortest F' in a DNF CDL problem, is describable in terms of the disjuncts of G'. An informal proof suggests that every disjunct in G' keeps negative examples out of the hypothesis space, but not each one is needed to cover the positive examples (S'). Therefore to simulate an Occam's Razor selection bias, we need only try combinations of G's disjuncts. Note that we do not yet say that it is identical to the F that would be produced by an approach employing an Information Gain metric (Quinlan, 1986) or Minimum Description Length principle (Quinlan, 1994)

Our agent's basic inductive learning algorithm is as follows:

1. Initialise two sets P and N according to whether we are generating G, S, F from examples, or using G and S to generate F.

2. Create an empty clause.

3. Add literal to the clause and remove from P and N training examples that do not follow from the clause.

4. Repeat 3 until N is empty

5. Save the clause, then reset N and repeat from 2 until P is empty.

6. Return the clauses.

This is basically the Separate and Conquer approach employed by FOIL and FOCL. However, the difference lies in how P and N are represented. If we are generating G, then N is set to the negative examples available, whilst P is initially defined intensionally as a set of size $sizeof(all\text{-}true) - sizeof(N)$. The reverse is true if we are generating S. We maintain P (or N) by calculating $sizeof(current\ clause)$. If we are generating F (from examples), then P and N are set to the examples. If we are generating F from G and S, then we set P to be S' and N to be the complement of G'.

We implement the function $sizeof(X)$ based on the fact we know the concept description language contains a finite set of literals. Consider a sentence consisting of a disjunction D of conjunctions $C_1..C_n$ of literals $L_1..L_n$. Thus $sizeof(X)$ is defined as:

$$sizeof(D) = (\Sigma_{C \in D}\, sizeof(C)) - sizeof(\bigcap_{C \in D} C)$$

$$sizeof(C) = \frac{1}{\prod_{L \in C} sizeof(L)} \prod_{1}^{n} sizeof(Ln)$$

$$sizeof(L) = |values\_of\_L|$$

3.1 Initial Observations

For small languages, the procedure works well. However it scales poorly. An analysis (prompted by initial experiments) of $sizeof(X)$ suggests it is NP-Hard. This is based on the observation that the algorithm is attempting to count the number of examples that would satisfy the G or S description. This is effectively the SAT problem (Floyd and Beigel, 1994) which is known to be NP-Complete. This is a disappointing result, but should have been expected. However, it is not grounds for giving up the approach.

Firstly, learning descriptions that are 3-DNF and above is known to be NP-Complete. Therefore we should not expect distribution to ameliorate the problem.

Secondly, there are a number of ways to deal with our problem:

1. Optimise the *sizeof(X)* algorithm so that is as efficient as possible, given the class it is in. An example of this would be to use bit vectors. Using this approach we have moved from being able to handle 4 binary attributes to 16 binary attributes with a reasonable response time (< 1 minute, on a 100Mhz RISC chip).

2. Approximate the G and S descriptions and calculations of *sizeof(X)*. This approach might also be desirable in order to handle noise in the training examples.

3. Reduce the Concept Description Language to a formally less hard problem (for example, learning descriptions which are 1-DNF).

In terms of evaluating the approach we will adopt the principle that the initial calculations are done "off-line". Thus we will compare the cost to generate F' from G' and S', with the cost of generating F' from all the examples plus the cost of the transmission.

Note that this initial agent can only deal with a fixed concept description language and selection bias. We discuss this restriction in section 5. However despite currently being restricted to a fixed CDL, the bounds G and S may be used with other selection biases. However, such biases (or rather the algorithms that implement them) will have to be modified to use the intensional definition of G and S.

4. The Proposed KQML Operators

Michalski's model of inductive inference is as follows: induction is the process of tracing backwards the tautological rule of universal specialisation; abduction is the result of tracing backward domain rules. Michalski notes that the first type of inductive inference is unsound, whilst the second may be unsound, depending on the strength of the reverse implication. However, it is possible to cast both forms of inductive inference within the version space model. It is also possible to place deductive inference within the version space model (see the previous section).

So our first modification to Michalski's theory is to replace his model of inference with the version space model. The second step is to make the operators refer to operations over version spaces. The final step is to cast them as KQML performatives that can be sent as requests between agents.

The second step is straightforward. In Michalski's terminology, the reference set defines the training examples (either intensionally or extensionally), and the descriptor set describes the representational biases. Michalski does not have a specific term for the selection bias, but offers the appropriate operator.

The final step is to define the KQML performatives in terms of the operators, their version space parameters. We also have to formally specify the semantics of these performatives, although currently, they are informally stated. An agent is then able to send and receive requests for inductive and deductive inferences. They are able to integrate the replies (which are in terms of hypotheses bounded by G and S) using version space intersection.

We are proposing the following basic extensions to KQML. This is not an exhaustive list of the new performatives, but it is a representative sample. The first performative we identify is *discriminate*. This is the basic request that specifies a version space. The actual version space can either be the set of all concepts, or the boundary sets. An agent receiving such a request can use its own knowledge to generate the version space, or it may "sub-contract" the task to other agents.

(DISCRIMINATE <goal-concept>,<positive-reference-set>,<negative-reference-set>, <descriptor-set>)

The next two pairs of performatives allow an agent to modify an inference request. *Associate* increases the membership of the original reference set, whilst *disassociate* (not shown), reduces it.

Abstract reduces the number of hypotheses that can appear in the version space. For example, by removing concepts that may be used, or constraining the language. *Concrete* (not shown) increases the number of different hypotheses.

(ASSOCIATE <goal-concept>, <reference-set>)

(ABSTRACT <goal-concept>, <descriptor-set>)

The *select* performative applies the selection bias to a given version space. Initially it will return a single hypothesis, but it may be possible to return an ordered set as well.

(SELECT <goal-concept>, <version-space>, <selection-bias>)

The final performative, *intersect,* is not one of Michalski's operators, but is required to integrate multiple version spaces.

(INTERSECT <goal-concept>, <version-space1>,..<version-spaceN>)

These performatives will be nested in usage. For example, an agent could send:

(REQUEST (SELECT g(?X) (INTERSECT g(?X) (DISCRIMINATE$_{agent1}$ g(?X))..)))

to an integrating agent in order to achieve the effect of distributed induction. Note that this formalism does not permit the use of version space and selected hypothesis triples of the form <G,F,S>. Thus an agent must ask (or tell) *Discriminate* and *Select* if it wishes to transmit the bounds and the hypothesis.

The parameters to these performative are informally defined as follows:

1. <goal-concept> : The name of the target concept.

2. <reference-set>: A definition of the examples. It could be an extensional set, or a set of knowledge that define the examples, etc. In many cases there will be two reference sets, representing positive and negative training examples.

3. `<descriptor-set>`: This defines the representational biases. As the representational bias is a language, we assume we will specify this using a grammar of an appropriate class. For a finite language a regular expression (Floyd and Beigel, 1994) will suffice.

4. `<version-space>`: The operators imply the generation of a version space, either enumerated fully, or defined as a boundary set. Therefore the *reply* to an operator request consists of a version space. It is also used as the input to and output from an *intersect* performative.

5. `<selection-bias>`: It is not yet certain how we can represent this. It is generally a procedural bias, therefore we could just name a procedure (e.g. *'most-general'*). However, it could also be specified declaratively (Smith, 1995).

These performatives and parameters are still under investigation, and will be more refined as we implement learning agents.

5. Discussion & Future Work

This is very much work in progress, and to date we have not fully expanded all the details. Nor have we evaluated the approach empirically. However, a number of issues have already been raised, which we will discuss here.

We believe our method subsumes two of the previous approaches to multi-agent learning: the voting mechanism can be thought of as a group of agents attempting to find at least one hypothesis that is contained in each of their version spaces for the problem; theory revision is equivalent to modifying a hypothesis so it is in the version space of the agent performing the revision. However, there are two problems with these methods. Firstly, it might not be possible to find one hypothesis consistent with all the version spaces even if one does exist, given that many algorithms use a hill climbing approach. Secondly, they result in the selection of a single hypothesis, which might be incorrect. The version space approach avoids both these problems. It is not certain whether our approach will have the efficiency of these general methods; or that of specialised distributed learning algorithms (or inherently incremental learners). However, we do believe that it is a general approach which should will allow inductive learning algorithms to be modified to used in agents. It is also perceivable that this approach can be specialised (for example, a neural network learning algorithm could send weight vectors that correspond to G and S), or can be generalised along the lines of (Smith 95) where all the biases are declarative.

The specific concerns so far are:

1. The approach adopted so far involves an NP-Complete algorithm.

2. We have not investigated how to handle noise and uncertainty, nor have investigated how to accommodate shifts in bias (Des Jardins and Gordon, 1995).

3. Our implementation is an attribute-value learner.

4. We do not have a mechanism for either integrating version spaces of different concept description languages or manipulating different reference or descriptor sets.

Future work will seek to address these issues in part. The first two points may well be jointly solved by approximating the version space boundary sets, using statistical information. The third point will be pursued by first taking the approach employed in LINUS (Lavrac and Dzeroski, 1994), and then examining the possibility of modifying an ILP algorithm such as GOLEM (Muggleton, 1992) to generate and use version spaces. Recent work by Nienhuys-Cheng and De Wolf (1996) suggests that certain classes of ILP problems are representable with boundary sets. We will not initially seek to address the fourth point because of the limited usage it would have, as version spaces of different languages can only be integrated on a lowest common denominator basis.

Acknowledgments

The authors would like to acknowledge funding by the UK Engineering and Physical Sciences Research Council for the initial phases of this research. They also acknowledge the support and encouragement of Mike Genesereth at Stanford University, and helpful comments from Nils Nilsson, Anna Patterson, Josefina Sierra, Lise Getoor, Ofer Matan, Ken Brown and Will Schuman.

Bibliography

P. Brazdil, M. Gams, S. Sian, L. Torgo, and W. Van de Velde (1991). Learning in Distributed Systems and Multi-Agent Environments, In *Proceedings of the European Working Session on Learning (EWSL91)*, Springer-Verlag, pages 424-439, Porto, Portugal.

P. Brazdil and L. Torgo (1990). Knowledge Acquisition via Knowledge Integration, in *Current Trends in AI*, B. Wielenga et al.(eds.), IOS Press, Amsterdam.

N. Cesa-Bianchi, Y. Freund, D. P. Helmbold, D. Haussler, R. E. Schapire, and M. K. Warmuth (1995). *How to Use Expert Advice*, Technical Report UCSC-CRL-95-19, University of California, Santa Cruz, CA.

P. K. Chan and S. J. Stolfo (1995). A Comparative Evaluation of Voting and Meta-Learning on Partitioned Data, In *Proceedings of the Twelfth International Conference on Machine Learning (ML95)*, Morgan-Kaufmann, pages 90-98, Lake Tahoe, CA.

W. Davies (1993). *ANIMALS, An Integrated Multi-Agent Learning System*, M.Sc. Thesis, Department of Computing Science, University of Aberdeen, UK.

M. Des Jardins and Diana F. Gordon (1995). Evaluation and Selection of Biases in Machine Learning, *Machine Learning*, 20:1-17.

T. Finin, J. Weber, G. Wiederhold, M. Geneserth, R. Fritzson, D. MacKay, J. McGuire, R. Pelavin, S. Shapiro, and C. Beck (1993). *Draft Specification of the KQML Agent-Communication Language*, Unpublished Draft.

R. W. Floyd and R. Beigel (1994). *The Language of Machines*, Computer Science Press, NY.

M. Gams (1989). New Measurements Highlight the Importance of Redundant Knowledge, In *Proceedings of the 4th European Working Session on Learning (EWSL89)*, Pitman-Morgan Kaufmann, pages 71-80, Montpellier, France.

M. Genesereth (1995). *Epilog for Lisp 2.0 Manual*. Epistemics Inc., Palo Alto, CA.

H. Hirsh (1989). *Incremental Version Space Merging: A General Framework for Concept Learning*, Ph.D. Thesis, Stanford University.

S. Jain and A. Sharma (1993). *Computational Limits on Team Identification of Languages*, Technical Report 9301, School of Computer Science and Engineering, University of New South Wales, Australia.

M. Kearns and H. S. Seung (1995). Learning from a Population of Hypotheses, *Machine Learning*, 18:255-276.

N. Lavrac and S. Dzeroski (1994). *Inductive Logic Programming: Techniques and Applications*, Ellis Horwood, Herts, UK.

R. S. Michalski (1993). Inferential Theory of Learning as a Conceptual Basis for Multistrategy Learning, *Machine Learning*, 11:111-151.

T. M. Mitchell (1978). *Version spaces: An Approach to Concept Learning*, Ph.D. Thesis, Stanford University.

T. M. Mitchell, R. M. Keller, and S. T. Kedar-Cabell (1986). Explanation-Based Generalization: A Unifying View, *Machine Learning*, 1:1-33.

R. J. Mooney and D. Ourston (1991). A Multistrategy Approach to Theory Refinement, In *Proceedings of the International Workshop on Multistrategy Learning*, pages 115-130, Harper's Ferry, WV.

S. Muggleton (1992). *Inductive Logic Programming*, Academic Press, London, UK.

S. H. Nienhuys-Cheng and R. De Wolf (1996). Least Generalizations and Greatest Specializations of Sets of Clauses, *Journal of Artificial Intelligence Research*, 4:341-363

M. Pazzani and D. Kibler (1992). The Utility of Knowledge in Inductive Learning, *Machine Learning*, 9: 57-94.

F. J. Provost and D. N. Hennessy (1995). Distributed Machine Learning: Scaling up with Coarse Grained Parallelism, In *Proceedings of the Second International*

Conference on Intelligent Systems for Molecular Biology (ISMB94), AAAI Press, pages 340-348, Stanford, CA.

J. R. Quinlan (1986). Induction of Decision Trees, *Machine Learning*, 1:81-106

J. R. Quinlan (1990). Learning Logical Definitions from Relations, *Machine Learning*, 5:239-266

J. R. Quinlan (1994). The Minimum Description Length Principle and Categorical Theories, In *Machine Learning, Proceedings of the 11th International Workshop (ML94)*, Morgan Kaufmann, pages 233-241, New Brunswick, NJ.

S. S. Sian (1991). *Learning in Distributed Artificial Intelligence Systems*, Ph.D. Thesis, Imperial College, UK.

B. Silver, W. Frawley, G. Iba, J. Vittal, and K. Bradford (1990). ILS: A Framework for Multi-Paradigmatic Learning, In *Machine Learning, Proceedings of the 7th International Workshop (ML90)*, Morgan Kaufmann, pages 348-356, Austin, TX.

B. D. Smith (1995). *Induction as Knowledge Integration*, Ph.D. Thesis, University of Southern California.

V. Svatek (1994). *Integration of Rules from Expert and Rules Discovered in Data*, Unpublished Draft, Prague University of Economics, Czech Republic.

P. E. Utgoff (1989). Incremental Induction of Decision Trees, *Machine Learning*, 4:161-186.

Addressee Learning and Message Interception for Communication Load Reduction in Multiple Robot Environments

Takuya Ohko[1], Kazuo Hiraki[2] and Yuichiro Anzai[1]

[1] Anzai Laboratory, Department of Computer Science, Keio University, 3-14-1,
Hiyoshi, Kohoku, Yokohama 223, Japan
E-mail: {ohko, anzai}@aa.cs.keio.ac.jp
WWW: http://www.aa.cs.keio.ac.jp/members/ohko/
[2] Presto / Electrotechnical Laboratory, MITI, 1-1-4, Umezono, Tsukuba 305, Japan
E-mail: khiraki@etl.go.jp

Abstract. This paper describes communication load reduction on task negotiation with Contract Net Protocol for multiple autonomous mobile robots. We have developed LEMMING[5, 6, 7], a task negotiation system with low communication load for multiple autonomous mobile robots. For controlling multiple robots, Contract Net Protocol(CNP)[10] is useful, but the broadcast of the Task Announcement messages on CNP tends to consume much communication load. In order to overcome this problem LEMMING learns proper addressees for the Task Announcement messages with Case-Based Reasoning[3] so as to suppress the broadcast. The learning method is called Addressee Learning. However the learning method causes inefficient task execution. An extension of LEMMING with Message Interception which enables the system to execute tasks more efficiently is reported.

1   Introduction

We launched a project called **PRIME** (Physically-grounded human-Robot-computer Interaction in Multiagent Environment) for understanding how multiple humans, robots and computers interact with each other, and for generating effective design methodologies for interaction[1]. In this context the reduction of the communication load is an important issue for cooperation among multiple autonomous mobile robots. Because it is difficult for multi-robot systems to use high-speed and wide-band communication lines, compared with desktop-computer systems, although it is important for the robots to communicate each other for coordinating their action even if they act independently.

To overcome the problem of communication load reduction among robots, we have developed a system called LEMMING[5, 6, 7] that learns to reduce the communication load for task negotiation using Contract Net Protocol(CNP)[10] with Case-Based Reasoning(CBR)[3]. However the learning method causes inefficient task execution.

This paper extends LEMMING with Message Interception to improve the efficiency of the task execution.

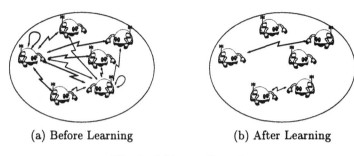

(a) Before Learning   (b) After Learning

Fig. 1. Addressee Learning

In general, the messages handled by CNP include Task Announcement messages, Bid messages, Award messages, and Report messages. It is observed that a high proportion of the communication load comes from the Task Announcement broadcasting. To reduce this load, Smith proposes the *focused addressing*[10]. But the knowledge for the focused addressing must be provided by human experts in advance. Thus, we proposes a technique to let robots acquire such knowledge by CBR.

In previous approaches, like CNET[10], the information gathered from a contract is not fully utilized and is disposed after each negotiation. Thus, this paper presents an attempt to extract useful knowledge from the messages in order to reduce the total number of message exchanges. CBR is employed to infer the most suitable robot to carry out the task specified in a particular Task Announcement message. This technique of learning with CBR is called *Addressee Learning*. LEMMING, residing in each robot, is developed to facilitate the task negotiation with CNP and Addressee Learning. LEMMING learns to reduce the communication load as shown in Fig. 1.

The suppression of the broadcast, however, causes the decrease of the efficiency of the task execution. Thus, we employ *Message Interception*. Message Interception allows robots to receive Task Announcement messages and to return Bid messages even if the robots are not the addressee of the Task Announcement messages. Message Interception improves the efficiency of the task execution because the manager of the task negotiation may gather better bids.

The performance of LEMMING is evaluated in a simulated multi-robot environment, where requests on robots consist of serving various dishes. Each robot has different dish. Thus, on receiving a request for service, a robot is required to negotiate a task with its peer robots if it does not have the requested dish. In this paper, the communication load for a message is assumed as the number of the robots addressed by the message. This evaluation is focuses on the reduction of the communication load and the waiting time for the dish.

Section 2 describes the architecture and the process of LEMMING. Section 3 proposes the Message Interception for extending LEMMING. Section 4 shows a sample environment in which mobile robots carry dishes to human guests. Section 5 evaluates LEMMING and that with Message Interception on both the communication load and the efficiency of the task execution in the sample environment. Section 6 introduces our studies on LEMMING and related work for reducing the communication load. Section 7 concludes this paper.

2 LEMMING: A Task Negotiation System with Low Communication Load

LEMMING is a task negotiation system with the low communication load for multiple autonomous mobile robots. LEMMING negotiates tasks with CNP. LEMMING reduces the communication load adaptively by Addressee Learning, a learning method to reason proper addressees for Task Announcement messages.

2.1 Architecture of LEMMING

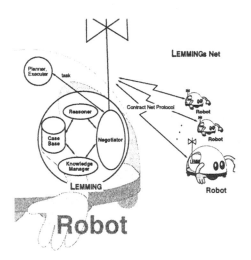

Fig. 2. Architecture of LEMMING

Each LEMMING controls a robot for negotiating tasks with Contract Net Protocol. Such robots with LEMMING composes LEMMINGs Net. LEMMING consists of Negotiator, Reasoner, Knowledge Manager, and Case Base (Fig. 2). Negotiator negotiates tasks with CNP. Reasoner reasons proper addressees for Task Announcement messages with Addressee Learning. Knowledge Manager manages cases and status of the negotiation process in Case Base.

2.2 Contract Net Protocol

Contract Net Protocol(CNP)[10] is a useful method for the task negotiation among multiple robots. CNP consists of Task Announcement message, Bid message, Award message, and Report message. When a task distributer called *manager* wants to allocate a task, the manager generates and broadcasts Task Announcement message for the task. The robot receiving the message evaluates it, and if the robot can execute the task, the robot evaluates the performance for the task execution, and returns Bid message with the performance. The manager selects the bid with the best performance, and appoints the sender of the bid as a *contractor*. The manager sends Award message to the contractor to execute the task. On finishing the task, the contractor returns Report message to the manager.

CNP consumes much communication load since Task Announcement message is always broadcast. To reduce the broadcast, *focused addressing*[10] is proposed to send Task Announcement message only to the proper addressees. But ordinary focused addressing needs the addressees fixed in advance. Then we proposed Addressee Learning so as to let robots learn the addressees adaptively by CBR.

2.3 Case-Based Reasoning

Case-Based Reasoning(CBR)[3] is a learning method to solve a problem with past cases (Fig. 3). A case consists of a problem and its solution. When a reasoning system is input a new problem, it searches the cases which include the similar problems to the new problem, and makes an answer of the cases. The problem and the answer are gathered and stored as a case to the Case Base incrementally. A case can be expressed with raw informations so that it can be gathered easily, although the size of Case Base tends to get larger according to the experience of the CBR system. Each case is independent with other cases so that the case is easy to be transferred to another Case Base and to be forgotten. Moreover the good aspect of the characteristic for human maintainer is that a case can be given, read, and corrected by human.

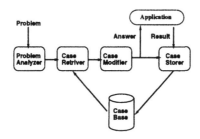

Fig. 3. Overview of Case-Based Reasoning

2.4 Addressee Learning

Addressee Learning is a learning method to reason proper addressees for Task Announcement messages with CBR. On a general task negotiation system with

CNP, a manager broadcasts Task Announcement message. But the broadcast tends to consume much communication load. Thus, LEMMING suppresses the broadcast with Addressee Learning. The learning policy depends on that good robots for past similar cases may also be good now. Addressee Learning stores cases incrementally, and reasons proper addressees for the task announcement of a new case with past similar cases. Moreover, Addressee Learning can deal with immigration and emigration, and capacity change of the robots.

2.5 Data Description

Task Description: Task T is a list of attribute-value pairs described as follows:

$$T = (A_1V_1, A_2V_2, \ldots, A_{m-1}V_{m-1}, \texttt{taskid}:taskid) \tag{1}$$

Where $A_k(k = 1 \ldots (m-1))$ is the kth attribute, and $V_k(k = 1 \ldots (m-1))$ is the value corresponding to A_k. The pair of taskid: and $taskid$ is an identifier of the task. The identifier should be added when the task is generated.

Message Description: Figure 4 shows the descriptions of the messages for the task negotiation. Task Announcement message, Bid message, Nack message, Award message, Report message are used for each contract.

Task Announcement message: (TYPE:ANNOUNCE, CONTRACT:*contract*, FROM:*from*, TO:*to*, TASK:*task*, BID-SPECIFICATION:*bid-specification*, ELIGIBILITY-SPECIFICATION:*eligibility-specification*, EXPIRATION-TIME:*expiration-time*)

Bid message: (TYPE:BID, CONTRACT:*contract*, FROM:*from*, TO:*to*, ESTIMATED-PERFORMANCE:*estimated-performance*)

Award message:
(TYPE:AWARD, CONTRACT:*contract*, FROM:*from*, TO:*to*, TASK:*task*, REPORT-SPECIFICATION:*report-specification*, RESULT-SPECIFICATION:*result-specification*)

Nack message: (TYPE:NACK, CONTRACT:*contract*, FROM:*from*, TO:*to*, RESULT:*result*)

Report message: (TYPE:REPORT, CONTRACT:*contract*, FROM:*from*, TO:*to*, RESULT:*result*, ACTUAL-PERFORMANCE:*actual-performance*)

Fig. 4. Message Description

contract is an identifier of the negotiation corresponded to by the message. *from* shows the sender of the message. *to* shows the addressee of the message. *task* is a task to be executed. *bid-specification* is the formula to estimate the performance for the task. *eligibility-specification* is a condition whether the receiver can return a bid or not. *expiration-time* is a time the manager waits for the bids. *estimated-performance* is a scalar value of the *estimated* performance for the task. *report-specification* is similar to the *bid-specification*, but it should calculate *actual* performance. *result-specification* is the formula for calculating result.

result is the result of the task execution. *actual-performance* is a scalar value which represents an actual performance for the task execution of the contractor at that time.

Case: A case is described in Fig. 5.

case: (CASEID:*case-id*,
    TASK:*task*, ANNOUNCE:*announce-list*, BID:*bid-list*, AWARD:*award-list*, NACK:*nack-list*, REPORT:*report-list*, GENTIME:*gentime*, RCOUNT:*RCount*, STATUS:*status*)

Fig. 5. Description of Case

case-id is the identifier of the case. *task* is a list of attribute-value pairs, and shows the task of the case. *announce-list, bid-list, award-list, nack-list, and report-list* are lists of Task Announcement messages, Bid messages, Award messages, and Report messages dealt with the case respectively. *gentime* shows what time the case is generated. *RCount* shows how many times the Reasoner called on the negotiation. *status* shows the status of the negotiation process. The values of *status* are announcing, executing, or finish.

2.6 Learning Process

Figure 6 shows the task negotiation process of LEMMING. This subsection focuses on the process of Addressee Learning in the task negotiation process.

Negotiator When a new task is input to LEMMING, Negotiator starts to negotiate the task with CNP. At first, the Negotiator calls Reasoner to decide which robot to send the Task Announcement message. Then the Negotiator sends the Task Announcement message to the selected robots. [1] If the Reasoner can not reason such robots or the Negotiator could not get any bid at the announcement, the Negotiator broadcasts the Task Announcement message.

Then, the Negotiator as a manager waits for the bids until all the bids from the addressees are returned or certain waiting-time(*expiration-time*) is passed. If no bid returned, the Negotiator calls the Reasoner again for re-announcing. When any bids returned, the Negotiator decides the best bidder as a contractor whose bid message has the best performance, and allocates the task to the contractor. When the contractor can not execute the task, it returns Nack message to let the manager re-announce. When the contractor finished the task, it returns a Report message.

---

[1] This paper does not concern directed contract[7], mediation, and recommendation.

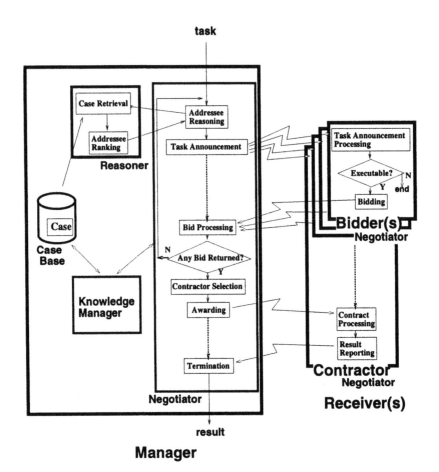

Fig. 6. Process of LEMMING

1. New task
2. Call Reasoner to decide addressees
3. Task announcement
   - (SUCCESS) Send to the addressees
   - (FAIL)
     - (1st) Broadcast
     - (ELSE) go to 7
4. Bidding
   - Waiting for the bids until all the bits from the addressees are returned or certain waiting-time(*expiration-time*) are passed
   - (NO BID) go to 2
5. Award
   - (the contractor can not execute the task and returns Nack message) go to 2
6. Report
7. Finish

Fig. 7. Process of Negotiator

Reasoner Reasoner searches such cases that include similar tasks to the input task. The function to evaluate the similarity is following:

$$Similarity(T_1, T_2) = \sum_{i=1}^{m} \sum_{j=1}^{n} Dist(A_{1i}, V_{1i}, A_{2j}, V_{2j}) \tag{2}$$

$$T_1 : (A_{11}V_{11}, \cdots, A_{1m}V_{1m}) \tag{3}$$
$$T_2 : (A_{21}V_{21}, \cdots, A_{2n}V_{2n}) \tag{4}$$

$$Dist(A_{1i}, V_{1i}, A_{2j}, V_{2j}) = W(A_{1i}, A_{2j})Equal(V_{1i}, V_{2j}) \tag{5}$$

$$W(A_{1i}, A_{2j}) = \begin{cases} 1 & \text{if } A_{1i} = A_{2j} \\ 0 & \text{otherwise} \end{cases} \tag{6}$$

$$Equal(X_1, X_2) = \begin{cases} 1 & \text{if } X_1 = X_2 \\ 0 & \text{otherwise} \end{cases} \tag{7}$$

Where T_1 and T_2 are the tasks to evaluate similarity. According to the similarities, the Reasoner sorts the cases, and selects similar cases with a threshold. The threshold changes according to the loop-counter($RCount$) which means how many times the Reasoner is called for the task. And the Reasoner calculates *"suitabilities"* of the robots from the similar cases:

$$suit(r) = average_{i,j}perf(m_r(Sc_i, j)) \tag{8}$$

$suit(r)$ is the suitability of robot r. Sc_i is a ith similar case. $m_r(Sc_i, j)$ is a jth Bid/Nack/Report message from robot r in Sc_i. $perf(m_r(Sc_i, j))$ is a performance value of $m_r(Sc_i, j)$. The suitability is an average performance of robot r extracted from the Bid, Nack, and Report messages in the similar cases. Then the Reasoner picks up some best robots according to the suitabilities, and returns a list of the robot-suitability pairs to the Negotiator.

When the Reasoner can not reason any addressee or the $RCount$ is over certain loop-limit, the reasoning is determined as failure.

Knowledge Manager Knowledge Manager makes cases. Reasoner refers the cases for Addressee Learning. And Negotiator also refers the cases to know the status of the negotiation processes. But the more the number of cases increase, the more the cost of the case retrieval and the space for holding the cases increase. Thus, the Knowledge Manager commits forgetting the oldest and finished cases if the number of the cases is over its limit.

3　Message Interception

This paper proposes Message Interception which improves the efficiency of the task execution by eavesdropping on the Task Announcement message for other robots.

3.1 Localization and Message Leakage

To utilize communication resources, there are many methods: a)reduction of number of messages, b)compression of the contents of messages, and c)localization of the communication range of messages. In multiple mobile robot environments, as the communication media, wireless communication is preferred. And the localization methods of wireless communication have 1)*cellular localization* and 2)*beam localization*.

Cellular localization is the method to restrict the power for emitting message with non-directed media. Beam localization is the method to make the direction of message narrower. These methods restrict communication range to improve utility of communication resources. But even with these localization methods, the messagecan be eavesdropped on by other robots. For example, Fig. 8 shows a message-leakage situation on cellular localization. In this situation, Robot C can eavesdrop on the message from Robot A because Robot C is nearer than Robot B. Figure 9 shows a message-leakage situation on beam localization. In this situation, Robot D can eavesdrop on the message from Robot A because the direction of Robot D is almost the same as that of Robot B.

Furthermore, as Fig. 10 shows, some localization need mediator-robots which mediate messages among the robots in different communication areas. The mediator-robots can also eavesdrop on the messages. When Robot E sends a message to Robot G, the mediator: Robot F can also overhear the message.

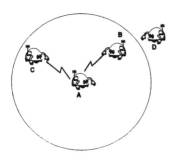
Fig. 8. Message Leakage Type (Cellular)

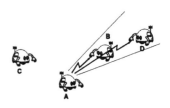
Fig. 9. Message Leakage Type (Beam)

Fig. 10. Message Leakage Type (Mediation)

3.2   Extension of Message Interception

This paper proposes Message Interception to utilize these message leakage to extend LEMMING so that the manager of the task negotiation can gather better bids to improve the learning speed and the efficiency of the task execution, although these message leakage causes the problems of information security and packet interference.

The extension of Message Interception to LEMMING is:

1. Negotiator keeps all the messages which are received correctly even if they are not addressed to the robot of the Negotiator.
2. If the type of the message is Task Announcement message, the Negotiator processes it for bidding.
3. The other messages which are not addressed to the robot are ignored.

This extension shows that high leakage rate is good for Message Interception. For example, the number of Bid message with that each message leaks to all the robots is almost the same as the number of Bid message with that each message is broadcast. But message leakage and broadcast are different in their controllability: leaked message is rarely intended to be, while broadcast message is often intended to be. Then the high leakage rate tends to be more harmful for the task negotiation than broadcast since not only Task Announcement message but also the other types of the messages always leak without control. Therefore it is important to reduce message leakage rate

4   Dinner Environment

This section decides "Dinner Environment" (Fig. 11), a sample environment for evaluating LEMMING.

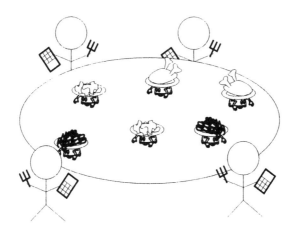

Fig. 11. Dinner Environment

Mobile robots carry dishes, and they know what they carry. The positions of the guests are situated, and each robot knows which guest is where.

A guest selects a dish, and inputs a task sequence with a handy remote controller to a randomly selected robot. The robot negotiates the task to the robot which is the fastest to carry the dish to the guest.

The task is expressed as a list of attribute-value pairs. For example, a task: "Serve Chicken to Guest G0" is described as (command:serve dish:Chicken to:G0 taskid:T0-1).

Definition of Performance The definition of performance is following:

$$Performance(R, t, T) = \begin{cases} \max(0, C - (EndTime(T) - t)) & (R \text{ can execute } T) \\ -1 & (R \text{ can not execute } T) \end{cases} \quad (9)$$

T is the task. t is the current time. r is the robot which evaluate the performance of T. $EndTime(T)$ is the time when R finished T. C is a constant which is enough large to $(EndTime(T) - t)$. We define that larger performance is better, and that (-1) means that R cannot execute T.

Definition of Communication Load The definition of the communication load is following:

$$ComLoad(M) = \begin{cases} NumAddr(M) & \text{(multicast or point-to-point)} \\ AllAddr & \text{(broadcast)} \end{cases} \quad (10)$$

M is the message to calculate the communication load. $NumAddr(M)$ means the number of the robots addressed by M. $AllAddr$ means the number of the robots in the environment.

5 Evaluation

There are six mobile-robots (R0, R1, R2, R3, R4, R5) and four human guests (G0, G1, G2, G3) in the environment. The speed of the robots is the same. Each robot serves dish. There are three kinds of dishes (Chicken, Salad, Pasta). The robots know what they carry, but do not know the capacity of the other robots at first.

- Number of Robots: 6
  (R0,R1,R2,R3,R4,R5)
- Kinds of Dishes: 3
  (Chicken,Salad,Pasta)
- Number of Human Guests: 4
  (G0,G1,G2,G3)

The values of the task are selected randomly, but the value of the taskid is generated as to be distinguishable from other values of taskid. The task is commanded to a randomly selected robot.

Each robot carries different dish(Table 1) : robot R0 and R1 carry Chicken, R2 and R3 carry Salad, and R4 and R5 carry Pasta. Thus, on receiving a request for service, a robot is required to negotiate a task with its peer robots.

Table 1. Relation of Robots-Dishes

Robot	Dish
R0,R1	Chicken
R2,R3	Salad
R4,R5	Pasta

The evaluation focuses on the reduction of the communication load and the waiting time for the dish. Thus, we extract the communication load and the waiting time per task by each robot in each simulation-loop; and show their average as the result of the evaluation. A simulation is finished when each robot has negotiated 40 tasks. The number of the simulation-loop is 30. The unit of waiting time is *clock*. It takes about 10 clocks for the robots to move from side to side on the table. The first Task Announcement message for each task is sent to the best suitability robot only. The communication medium is properly multiplied so that messages should not be conflicted nor delayed. All the broadcast message are always successfully delivered to all the robots.

As comparison systems with LEMMING, *broadcast-based contract net system* is evaluated. The broadcast-based contract net system always broadcasts Task Announcement messages. And the LEMMING with Message Interception is also evaluated. To evaluate Message Interception, we assumes that a message is received by the robots which are not intended as the addressees at the percentage of 20 or 60. (These high leakage rate are to make the results easy to show.)

– Comparison systems

- broadcast-based contract net system
- LEMMING
- LEMMING with Message Interception (20%)
- LEMMING with Message Interception (60%)

5.1 Communication Load

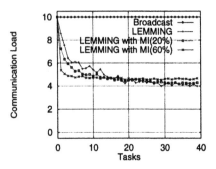

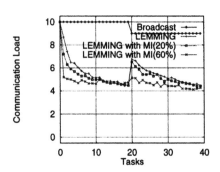

Fig. 12. Communication Load   **Fig. 13.** Communication Load (run out)

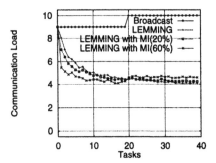

Fig. 14. Communication Load (filled)

Table 2. Relation of Robots-Dishes(dishes of R1,R3,R5 are empty)

Robot	Dish
R0	Chicken
R2	Salad
R4	Pasta
R1,R3,R5	empty

Figure 12 shows the average total communication load. Figure 13 also shows
the average total communication load, but the dishes of R1, R3, and R5 become

empty(Table 2) when each robot negotiates 20 tasks. Figure 14 also shows the average total communication load, but the dishes of R1, R3, and R5 are empty before each robot negotiates 20 tasks.

These are plotted against the number of the experienced tasks with broadcast-based contract net system, LEMMING, and LEMMINGs with Message Interception.

Table 3. Total Communication Load(After Learning)

	Ann.	Bid	Award	Nack	Rep.	total
Broadcast	6	2	1	0	1	10
LEMMING	1	1	1	0	1	4
LEMMING(MI20%)	1	1.2	1	0	1	4.2
LEMMING(MI60%)	1	1.6	1	0	1	4.6

Table 3 shows total communication load after learning. The calculation of the communication load is:

- The load of broadcast-based system consists of Task Announcement message× ⟨*AllAddr*⟩, Bid message×⟨*NumberOfTheRobotsWithProperDishes*⟩, Award message×1, and Report message×1.

- The load of learned LEMMING consists of Task Announcement message×1, Bid message×1, Award message×1, and Report message×1.

- The load of learned LEMMING with Message Interception consists of Task Announcement message×1, Bid message×(⟨*NumberOfTheProperAddressees*⟩+ ⟨*MessageLeakageRate*⟩ × (⟨*NumberOfTheRobotsWithProperDishes*⟩− ⟨*NumberOfTheProperAddressees*⟩)), Award message×1, and Report message×1.

Thus, Figure 12 suggests that LEMMING can reduce the communication load for the task negotiation. With Message Interception, learning speed is improved, but the communication load after learning becomes higher than that of normal LEMMING because of the increase of the number of bids. And the higher the message leakage rate is, the higher the increase rate of the number of bids is.

Figure 13 shows the adaptability of LEMMING. This figure suggests that the adaptability of LEMMING is also improved with Message Interception. And the higher the message leakage rate is, the more the adaptability is.

5.2 Efficiency of Task Execution

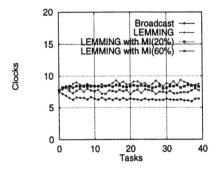

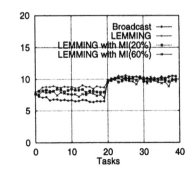

Fig. 15. Efficiency of Task Execution

Fig. 16. Efficiency of Task Execution(run out)

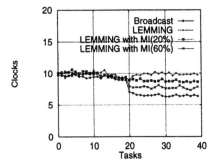

Fig. 17. Efficiency of Task Execution(filled)

Figure 15 shows the efficiency of the task execution(average waiting time). Figure 16 also shows the efficiency of the task execution(average waiting time), but the dishes of R1, R3, and R5 become empty when each robot negotiates 20 tasks. Figure 16 also shows the efficiency of the task execution(average waiting time), but the dishes of R1, R3, and R5 are empty before each robot negotiates 20 tasks.

These are plotted against the number of the experienced tasks with broadcast-based contract net system, LEMMING, and LEMMINGs with Message Interception.

These figures suggest that the efficiency of the task execution of LEMMING is lower than that of broadcast-based system, but thats of LEMMINGs with Message Interception are improved especially in Fig. 17.

In the later area of Fig. 16 and the first area of Fig. 17, there is only one robot which carries each dish, so the waiting times of the systems are the same.

On the contrary, in the first area of Fig. 16 and the latter area of Fig. 17, there are two robots which carry each dish, but the waiting times of the systems are not the same except that of the broadcast-based system. The reason is that the situation in the latter area of Fig. 17 is almost the same situation as new robots join to the environment, and the learned LEMMING can not utilize the new robots effectively. The broadcast-based system and LEMMINGs with Message Interception can utilize the new robots so that the waiting-times of the systems are decreased.

These results suggest that utilizing message leakage with Message Interception is effective for improving the speed of the learning and the efficiency of the task execution, but increases the communication load as shown in the previous subsection. Although these results also seem to suggests that intentional message leakage is effective, the message leakage itself is harmful for communication; thus, the message leakage should be restricted.

6   Related Work

Primitive LEMMING[5, 6] has been extended directed contract[7], forgetting[7], improving similarity function, recommendation, and so on. LEMMING can not only reduce the communication load but also make human-robot interface simple and friendly[4].

Ramamritham and Stankovic apply focused addressing on multiprocessor systems[8]. But their method needs computation time for each task. Then it is difficult for their method to apply for heterogeneous robots in dynamically changing environments. Shaw and Whinston use Classifier System and GA[9]. But the expression of the knowledge for focused addressing on their method is difficult to be read by human. Garland and Alterman use Collective Memory which they propose to make agents learn cooperative activity[2].

Suzuki et al. use eavesdropping for monitoring multi-robot system without special message for the monitoring[11].

7   Conclusion

This paper described the communication load reduction on task negotiation with Contract Net Protocol(CNP) for multiple autonomous mobile robots.

We have developed LEMMING, a task negotiation system with the low communication load for multiple autonomous mobile robots. LEMMING learns proper addressees for the Task Announcement messages on CNP with Case-Based Reasoning(CBR) so as to suppress the broadcast. The learning method is called Addressee Learning. However the learning method causes inefficient task execution. The paper reported an extension of LEMMING with Message Interception so that the system executes tasks more efficiently. The performance of LEMMING and the extension of Message Interception was evaluated in a simulated multi-robot environment. The evaluation shows that LEMMING can reduce the communication load adaptively, and that Message Interception improves the speed of the learning and the efficiency of the task execution. Message Interception is also effective incorporating new robots into the environment.

Acknowledgments

This research is supported by JSPS Research Fellowships for Young Scientists.

References

1. Yuichiro Anzai. Human-robot-computer interaction: A new paradigm of research in robotics. *Advanced Robotics*, 8(4):357–383, 8 1994.
2. Andrew Garland and Richard Alterman. Multiagent learning through collective memory. In Sandip Sen, editor, *Working Notes for the AAAI Symposium on Adaptation, Co-evolution and Learning in Multiagent Systems*, Stanford University, CA, March 1996.
3. Janet L. Kolodner. *Case-Based Reasoning*. Morgan Kaufmann, 1993.
4. Takuya Ohko and Yuichiro Anzai. A user friendly interface for task assignment on multiple robots by contract net negotiation with case-based reasoning. In *Abridged Proc. of the 6th International Conference on Human-Computer Interaction(HCI'95)*, page 12, 1995.
5. Takuya Ohko, Kazuo Hiraki, and Yuichiro Anzai. LEMMING: A learning system for multi-robot environments. In *Proc. of the 1993 IEEE/RSJ International Conference on Intelligent Robots and Systems, Yokohama, Japan(IROS'93)*, volume 2, pages 1141–1146, 1993.
6. Takuya Ohko, Kazuo Hiraki, and Yuichiro Anzai. Learning to reduce communication cost on task negotiation among multiple robots. In *Working notes of the IJCAI-95 Workshop on Adaptation and Learning in Multiagent Systems (also in Adaption and Learning in Multi-Agent Systems, Lecture Notes in Artificial Intelligence 1042, pp.177-190, Springer-Verlag)*, pages 65–70, 1995.
7. Takuya Ohko, Kazuo Hiraki, and Yuichiro Anzai. Reducing communication load on contract net by case-based reasoning – extension with directed contract and forgetting –. In *Proc. of 2nd International Conference on Multiagent Systems(ICMAS-96)*, pages 244–251, 1996.
8. Krithivasan Ramamritham and John A. Stankovic. Dynamic task scheduling in hard real-time distributed systems. *IEEE Software*, pages 65–75, 7 1984.
9. Michael J. Shaw and Andrew B. Whinston. Learning and adaptation in distributed artificial intelligence systems. In Les Gasser and Michael N. Huhns, editors, *Distributed Artificial Intelligence Volume II*, pages 119–137. Morgan Kaufmann, 1989.
10. Reid G. Smith. The contract net protocol: High-level communication and control in a distributed problem solver. *IEEE Transactions on Computers*, C-29(12):357–366, 1980.
11. Tsuyoshi Suzuki, Kazutaka Yokota, Hajime Asama, Hayato Kaetsu, and Isao Endo. Cooperation between the human operator and the multi-agent robotic system – evaluation of agent monitoring methods for the human interface system –. In *Proc. of the 1995 IEEE/RSJ International Conference on Intelligent Robots and Systems (IROS'95)*, pages 206–211, 1995.

Learning and Communication in Multi-Agent Systems

Holger Friedrich, Michael Kaiser*, Oliver Rogalla , and Rüdiger Dillmann

University of Karlsruhe, Institute for Real-Time Computer Systems & Robotics,
D-76128 Karlsruhe, Germany

Abstract. This paper discusses the significance of communication between individual agents that are embedded into learning Multi-Agent
Systems. For several **learning tasks** occurring within a Multi-Agent
System, communication activities are investigated and the need for a
mutual understanding of agents participating in the learning process is
made explicit. Thus, the need for a **common ontology** to exchange
learning-related information is shown. Building this ontology is an additional learning task that is not only extremely important, but also
extremely difficult. We propose a solution that is motivated by the human ability to understand each other even in the absence of a common
language by using alternative communication channels, such as gestures.
We show some results for the task of cooperative material handling by
several manipulators.

1 Introduction

Learning in Multi-Agent Systems has become a major research field within
Distributed Artificial Intelligence and Machine Learning [31, 27]. It is motivated
by the insight that it is impossible to determine **a-priori** the complete knowledge
that must exist within each component of a distributed, heterogeneous system
in order to allow satisfactory performance of that system. Especially if we want
to exploit the potential of modularity, such that it is possible for individual
agents to join and leave the Multi-Agent System, there's a constant need for
the acquisition of new and the adaption of already existing knowledge, i.e., for
learning.

Within this setting, different kinds of learning tasks must be investigated, such
as "traditional" single agent learning tasks, learning in teams, learning to act
within a team, and learning to coordinate other agents. To solve any of these
tasks, the existence of appropriate **information** that can be **communicated**
to the learning agents is of primary importance. Since any learning agent may
represent its individual knowledge by means of a formalism that will – in the
general case – not to be known to other agents, a common, agent-independent
language must be used for communication during task negotiation, cooperative

---

* Now with ABB Corporate Research Ltd., CHCRC.C, CH-5405 Baden-Dättwil,
Switzerland

task execution, and, especially, learning. More important, the communicating agents must share a common **ontology**, i.e., any agent receiving a message must understand it exactly in the way it was meant. Especially during communication between pupil and teacher, this is a crucial point.

However, when designing such a language we experience the same dilemma that initially motivated the use of learning: If we have a known, agent-independent "area" that we want to describe by a language, both syntax and semantics of that language can be defined a-priori. KQML [8] is a good example of such a language that has been designed specifically to facilitate **communication** in a content-independent manner. If the area or function we want to describe is not a-priori given or sufficiently complex, a general-purpose language such as KIF [9] is a reasonable choice. Nevertheless, the problem of "meaning" still exists: Imagine an agent a joins a team and wants to inform the members of the team that it has a capability c. If the team members know c, and the new agent and the team share the same symbol for c, that's no problem. In the general case, however, c may be completely new for the team, the symbol used by a to describe c may mean nothing (or something different) to the team, and the symbols used by the team may mean nothing to a, such that it cannot translate its internal representation appropriately. All these things can happen despite a common communication language and a common content language. This problem, for which currently solutions are being sought via tools for creating, accessing, and maintaining ontologies [7], results therefore in another learning task, the task **to learn the meaning of symbols.**

Throughout this paper, we will analyze the learning tasks existing within a Multi-Agent System with respect to their requirements regarding the communication between the agents involved. We'll show that understanding each other is of primary importance during learning, and what understanding really means with respect to a specific learning task and scenario. Finally, we will present an approach to learn the meaning of symbols in a cooperative way, which exploits "non-verbalizable" knowledge within several agents and the language-related competence of a knowledgeable agent, should the latter exist. This approach will be illustrated by means of an example of cooperative handling of flexible material. To provide the formal basis for these investigations, we'll employ state space models, following previous work by Beer [3] and ourselves [15].

2 Modeling Multi-Agent Systems

2.1 Single agent model

Our model of an agent a is that of a **skilled subsystem.** A single agent is able to perform **competent** actions that are related to its locally (possibly internally) defined goal and facilitate goal-oriented state transitions. More specifically, an important component of an agent a is the agent's strategy C_a^g with

$$C_a^g(\boldsymbol{x}) = \boldsymbol{u}. \tag{1}$$

$C_a^g(\boldsymbol{x})$ determines the goal-oriented action $\boldsymbol{u}$ that the agent executes if it is in state $\boldsymbol{x}$. To actually perform such a competent action requires **perceptual** capabilities (to determine $\boldsymbol{x}$), **cognitive** capabilities (to calculate $\boldsymbol{u}$), and **effectual** capabilities (to execute or apply $\boldsymbol{u}$, respectively). The action itself may be a physical one, such as the motion of a robot, or some "intellectual" effort, such as a request for data or a planning step. Similarly, the state $\boldsymbol{x}$ may include physically measurable information about the current situation of the agent with respect to its environment as well as internal, "mental" state variables such as estimations of the validity of internal models etc.

2.2 Multi-Agent System model

The Multi-Agent System is also a **skilled** system. Its actions – being compositions of the actions of all agents in the system – should be competent with respect to a system-wide goal. Therefore, a Multi-Agent System is modeled by a strategy function

$$C^G(\boldsymbol{X}) = \boldsymbol{U}, \tag{2}$$

where G denotes the current global goal of activity that is being pursued, $\boldsymbol{X}$ is the system's state and $\boldsymbol{U}$ is the calculated action.

On the lowest level of abstraction, the state vector $\boldsymbol{X}$ of a Multi-Agent System is simply the collection of all state vectors $\boldsymbol{x}_1, \ldots, \boldsymbol{x}_n$ of all embedded agents $a_1, \ldots, a_n$. Also, its action vector $\boldsymbol{U}$ is built from the individual action vectors $\boldsymbol{u}_1, \ldots, \boldsymbol{u}_n$, such that

$$\boldsymbol{X} = (\boldsymbol{x}_1{}^T, \ldots, \boldsymbol{x}_n{}^T)^T$$

and

$$\boldsymbol{U} = (\boldsymbol{u}_1{}^T, \ldots, \boldsymbol{u}_n{}^T)^T,$$

respectively. Fig. 1 illustrates this situation for a system (a service robot) consisting of three agents, an active camera system C, a mobile platform P, and a manipulator M. However, in the general case a process of **abstraction** is necessary to maintain modularity and extendibility and to allow for coordinated actions. Abstraction, however, means to associate signals (state vectors, action vectors), clusters of signals or sequences of signals to **symbols** that have a common meaning for all agents within the system. In other words, abstraction is a process that is complementary to explanation required for learning the meaning of symbols, i.e., to explain the meaning of a symbol, we must reverse the process of abstraction. We'll exploit this duality later.

3 Learning and Communication in Multi-Agent Systems

The purpose of learning in a Multi-Agent System is to enhance the system by extending its general capabilities or by improving its performance with respect to a particular criterion. In terms of the Multi-Agent System model (represented via equation (2)), learning comprises to change an existing function $C^G(\boldsymbol{X})$ as well as to newly define functions $C^{G_n}(\boldsymbol{X})$ that enable the system to pursue a new

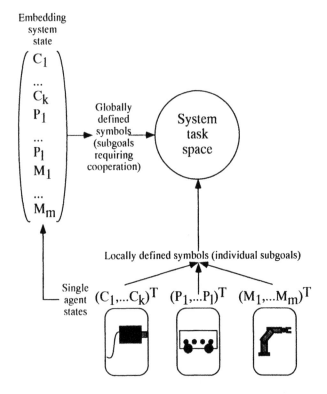

Fig. 1. Example for the definition of the system's state vector, action vector, and task space on the lowest level of abstraction, i.e., by means of local, single agent related subgoals. Cooperation related subgoals require abstraction.

goal G_n. Both tasks include several subtasks, such as learning in a single agent (isolated learning), learning to act as a team[2], building symbols from signals, and learning the meaning of those symbols.

3.1 Isolated learning

Learning within a single agent a means to alter the action u calculated via $C_a^g(x) = u$ from the state vector x or to build a new function $C_a^h(x) = u$ that enables the agent to contribute to a subgoal h.

In the first case, each learning agent obtains specific feedback that can be used to alter its action u. In the most simple case, the optimal action u^* is communicated to the agent, such that **incremental supervised learning** may take place. [24] describes such a situation for an agent (a robot) that learns directly from user demonstrations. As in adaptive control, agents may also receive an indication

---

[2] It should be noted that we use the term "learning to act as a team" instead of "team learning," in order to distinguish the "team learning" scenario (multiple agents trying to learn the same concept/language, as in [13]) from the situation considered here.

of the direction Δu into which to alter their actions, instead of an optimal action [16]. Another prototypical setting is that of **reinforcement learning,** in which the agent receives a possibly delayed reward r as feedback (see, for example, [20]) and alters its actions in order to maximize the reward. This kind of learning requires **exploration,** i.e., the systematic alteration of the calculated action in order to estimate the optimal action.

In the second case, learning agents must discover that the instruction they get from the teacher is related to a new subgoal h. Following that, they can build a function C_a^h from these instructions. This setting is typical for transferring skills from one agent (e.g., a human) to another (e.g., a robot) [15].

Communication issues in isolated learning In the isolated learning case, a single agent (the instructed agent or the "pupil") receives feedback from another agent. This agent may be an artificial one (a softbot), or a human supervisor. If learning takes place in a supervised manner, such that the given feedback consists of an optimal action or a quantitative indication of the error made by the agent, either

- the teacher must know the action space of the instructed agent and formulate the advice appropriately,

or

- the instructed agent must be able to map the teacher's advice onto its own action space.

Both requirements are not trivial, especially if agents should learn from other agents that are not structurally identical.

For learning on the basis of a scalar reward, the situation is very similar. Either

- the teacher must know the range of possible rewards used by the instructed agent (i.e., what is the "good" and "bad" in terms of the pupil),

or

- the instructed agent must know the mapping between the teacher's reward and its own range of rewards (what does the teacher mean by "good" and "bad").

In all cases, teacher or pupil must initiate the learning process. An important requirement is also that the teacher knows the limits of the instructed agent, since it makes no sense to try to teach an agent to go beyond its maximum capabilities (e.g., to try to position a robot with a higher precision than it is capable of). To enable the teacher to take care of this aspect requires the teacher to query these limits from an agent and to correctly interpret the agent's answer. Similarly, the instructed agent must understand the teacher's request and relate its capabilities properly to the task specified by the teacher. Summarizing, the following communication-related activities are parts of an isolated learning task:

1. Initiation of the learning process (activation of the pupil's "learning engine"), including communication of the learning context (the agent-related subgoal g or a new subgoal h) to the pupil.
2. Request a description of the pupil's limitations with respect to the current learning task.
3. Communication of the pupil's limitations to the teacher.
4. Communication of a target value $(u^*, \Delta u, r)$ from the teacher to the pupil.
5. If the teacher requires a model of the pupil, communication of information about success/failure of the learning process from the pupil to the teacher.

3.2 Learning to act as a team

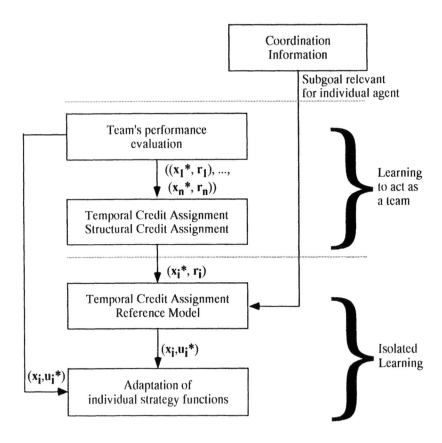

Fig. 2. Reduction of a task of learning to act as a team to isolated learning tasks in both the supervised learning and the reinforcement learning scenario.

While learning to act as a team, the feedback available for learning is not related to the performance of a single agent but to that of a team of agents. I.e., learning aims at improving the performance of a team with respect to a common criterion.

As in the isolated case, the setting that is easiest to handle is that of **incremental supervised learning.** Here, a teacher provides the optimal action that the team should take in each state. If team states and team actions can be directly mapped onto single agent states and actions, this scenario is equivalent to the one described in section 3.1, if also the function $C_a^g(\boldsymbol{x}, \boldsymbol{u})$ (the relevant subgoal) to be adapted is known for every single agent a. This is the case if we assume that task negotiation and agent coordination (the association of $g_{a_1}, \ldots, g_{a_n}$ to agents $a_1, \ldots, a_n$) have been done properly, i.e., based on symbols that correctly describe agent capabilities and constraints related to agent cooperation. As in the single agent case, agents a_i may also receive instructions to learn new functions $C_{a_i}^{h_i}(\boldsymbol{x})$ that are related to individual subgoals h_i or functions $C_{a_i}^h(\boldsymbol{x})$ related to team-specific subgoals h.

If only a scalar evaluation of the team's performance is available, the problem of **credit assignment** becomes evident. In contrast to isolated learning, which may incorporate the task of **temporal credit assignment,** i.e., the determination of those actions that are responsible for a delayed reward, learning to act as a team from feedback also involves a **structural credit assignment problem.** For each agent being a member of the team, its contribution (positive or negative) to the obtained reward must be determined. If agents are conscious about the performance of other members of the team, this process can be supported by requesting team members to provide individual evaluations. Another possibility is to temporarily assign the responsibility for choosing actions to one of the cooperating agents. This agent determines for each member of the team which action to perform, such that the learning task, exploration and credit assignment are reduced to a single agent task in higher-dimensional perception and action spaces [28].

Once credit assignment has taken place and subgoals g_{a_j} (or h_j or h, respectively) are known for any of the team members, team learning is reduced to several parallel steps of isolated learning (see Fig. 2 and, for example, [28]). It should, however, be noted that in some cases additional constraints are to be observed, since a single agent's environment (and, possibly, its target function) is continuously changing [26].

Communication issues in learning to act as a team When learning to act as a team, the teacher instructs or evaluates a group of agents that cooperate towards a common goal g. Here, the same communication requirements as in the isolated learning case exist (see section 3.1). In addition, an agent a must be able to identify which strategy C_a^g should be subject to the taught changes. To facilitate this identification step, either

- the teacher is able to define an appropriate subgoal g_i for each agent a_i in the team

or

- all instructed agents a_i are able to determine their respective subgoal g_i from the team-related goal g.

Consequently, agents must share a common terminology – symbols that describe team states, state-space trajectories, team actions, etc. Then, learning to act as a team comprises the following activities related to communication:

1. Negotiation for credit assignment (if supported).
2. Isolated learning for all team members.
3. If the teacher requires a model of the team, communication of information about success/failure of the learning process from each team member to the teacher.

3.3   Learning to communicate

The usefulness of communication depends on the ability of the communicating entities to understand each other. This is especially true for the communication between teacher and pupil, since the pupil performs self-modifications based on its understanding of the information obtained from the teacher.

In a Multi-Agent System, also an agent that coordinates a team of agents as well as agents cooperating in a team must be able to understand their respective counterparts – independent on the coordination/negotiation technique (see [29, 19] for examples) that is actually used. They must be able to understand what they are expected to do (or how they could contribute), and must be prepared to formulate their requests in an understandable manner.

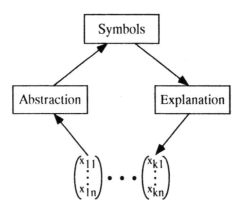

Fig. 3. Duality of abstraction and explanation.

To understand the meaning of symbols used for communication, these symbols must be *grounded* on the representation primitives of each agent – its state and its action vector [21, 11]. Such symbols can be general, such that they describe phenomena that can only be observed within a team or on the system level and are only partially understandable for most individual agents. They may also be specific, such that they represent, for example, a capability of a single agent. In both cases, however, a symbol is only useful if it is understood by at least two agents, such that it can be used for communication purposes. Consequently, it is not sufficient to provide means for building symbols (via abstraction), it must also be possible to explain the meaning of symbols to agents (Fig. 3).

3.4 Abstraction: From signals to symbols

Learning to map signals onto symbols adds a new dimension to the relationship between teacher and instructed agent. A new symbol affects all agents within the Multi-Agent System, since it requires possibly all of them to **extend their vocabulary.** The targeted construction and invention of a new symbol is an extremely difficult task, and the individual problems discussed in the following are in general not yet solved.

Nevertheless, we can identify three situations that require to build a new symbol:

1. The activities of an agent or a team of agents yield a system state (or state-space trajectory) that is considered to be characteristic or useful, i.e., that represents a subgoal.
2. For specifying a task, a new subgoal (to be pursued by a single agent or a team) must be defined that is not yet represented by a symbol.
3. An agent or a team develops a new capability (or a new agent with a new capability joins the Multi-Agent System) that is useful to know and to use when specifying and negotiating about tasks.

Whenever such a situation occurs (this can possibly only be detected by a knowledgeable agent, such as a human user), the symbol to be defined will be grounded on the current state (or state-space trajectory) of the system, the respective team, or the respective agent. In this sense, the symbol represents a concept that is to be learned in a supervised manner, for example by methods such as those described in [10, 18].

In addition, the Multi-Agent System's task space must be extended by the new subgoal. Task-specifying external agents and the Multi-Agent System itself must incorporate the new symbol into their task-description language. To explain the meaning of the new symbol to those agents that did not participate in the grounding process, this symbol must be translated to those agents, as described in the next section.

3.5 Explaining the meaning of symbols

The communication activities related to the task of symbol learning differ significantly from those occurring within other learning tasks. This is due to the fact that – following the creation of a new symbol – we can't simply use that symbol for self-explanation. Instead, we must **explain** what it means, i.e., we must circumscribe its meaning in already well-established terms.

To perform this task, two possibilities exist. First, we may be lucky and find an agent within the Multi-Agent System that understands both the symbol to be explained and the language of the agent that needs explanation. For example, a human supervisor will almost certainly be such a knowledgeable agent that may explain the meaning of the newly generated symbol to the learning agent, e.g., via direct implementation.

Second, it might be possible to choose another communication channel different from the verbal/textual one to explain the symbol. In analogy to humans, this channel may be something like the gestural-visual one. The meaning of a symbol is explained by demonstration, i.e., by performing the actions or establishing the state (or state-space trajectory) that is represented by the symbol. Since there are agents on whose states, state-space trajectories, or actions the symbol has initially been defined, these agents can literally show what the symbol means, while the agent needing explanation observers and analyzes this demonstration. In principle, this mechanism works in two directions: Agents within a Multi-Agent System can perform actions that are observed by an agent that has just joined the system, in order to explain the meaning of symbols used for communication to that agent. Also, a new agent can demonstrate the meaning of its own symbols to the members of a team or a Multi-Agent System, in order to make himself understandable to those.

Obviously, observing other agents' actions and matching the observations against one's internal representations is not a trivial task. However, many techniques developed within the framework of Programming by Demonstration [4] may prove useful to support it. After all, the idea to explain the meaning of symbols ("words") in a non-verbal manner is applied very successfully in everyday human communication.

3.6 Hidden states

In any of the scenarios described above, the hidden state problem may occur. Agents may not be capable to completely perceive all information that is necessary for them to uniquely identify a state. Also teams of agents may fail to do so. If no additional restrictions are imposed on the actual learning task, additional external information is mandatory to solve this problem. This information may be given by other agents, who may communicate their internal state variables or their perceptions to agents experiencing the hidden state problem (e.g., [22]). This allows the latter agents to distinguish internally equal states by means of additional external information, however for the price of increased state complexity. To automatically determine which information provided by other agents to choose in order to obtain a minimal unambiguous internal state description is a topic of current work.

4 Examples

In this section, we will present examples of how implicit mechanisms of explanation can be used to transfer knowledge to other agents. First, we shortly discuss transfer techniques for a 1:1 scenario. Afterwards, we present a detailed example of team learning.

4.1 Skill acquisition and the learning of individual symbols

One of the major goals in individual learning agents is to let the learning agent develop useful **skills** over time. In the case of robots, these skills are not only related to physical actions of the robot, but also to the coordination of activities, communication with humans, and active sensing. If we consider agents that learn such skills from other agents, both *skill acquisition from human demonstration* and *learning via imitation* are potentially useful methods.

On the one hand, human demonstrations can be used as the basis for learning both, *open-loop and closed-loop elementary skills*. The acquisition of open-loop skills is mostly focused on the reconstruction of trajectories from a sequence of demonstrated states (positions) [5], [30]. Systems supporting the acquisition of closed-loop elementary skills comprise acquisition techniques for manipulation tasks such as deburring [2] and assembly [15] as well as for vehicle control [23] and autonomous robot navigation [24], [14].

Learning new perceptive skills for object and landmark recognition can also take place on several system control levels. [1] presents an approach to learn sensor parameterizations from demonstrations. Learning active perception skills, i.e., the combination of actions and sensing for the purposes of object recognition, is the topic of work presented in [18].

Robot imitative mechanisms are relatively new, and learning by imitation has been demonstrated in the context of robot-robot interaction. In [12, 6], it was shown how a robot can learn to identify and handle the different corners of a maze by imitating the movements of a teacher robot. While the teacher robot navigates through the maze, the learner robot follows it and imitates its movements. However, while it is doing that, it associates the environmental state perceived through its sensors (i.e. the wall configuration) with the actions it is performing due to its imitation mechanism.

4.2 Learning to act as a team

In order to transfer knowledge to a team of agents, first a stream of instructions provided by a teacher has to be segmented into a sequence of symbols describing the task and, possibly, into individual instructions for each single agent. This segmentation can, for example, be performed by methods such as those presented in [17]. Afterwards, the whole task description is represented in a tree which consisting of several kinds of nodes (Fig. 4).

Symbol nodes represent a particular symbol in the MAS' language. The symbol itself is defined as a sequence of instructions. If an agent does not know the meaning of a symbol, it can be immediately provided with the associated instructions. On the lowest level, these instructions are given in terms of symbols denoting elementary skills. If an agent does not understand any of those symbols, skill acquisition as described in the previous section has to take place. For the purposes of this example, we assume that the participating agent possesses the required skills.

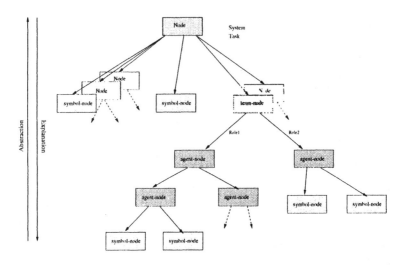

Fig. 4. Organization of symbols, team-tasks, and individual tasks in a tree.

Team nodes represent tasks which require cooperation between individual agents. Children of such nodes represent **roles** that can be filled by agents belonging to the team. Finally, **agent nodes** represent subtasks that have to be performed by a single agent.

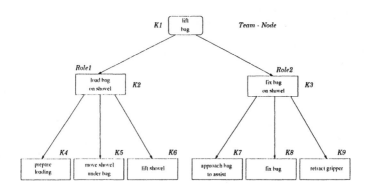

Fig. 5. Explanation tree for the subtask *lift-bag*.

Learning within this representation requires to explain roles and symbols and the meaning of a role within a subtask (represented as a symbol). Fig. 5 shows an instantiated tree for the explanation of a subtask within the *transfer-bag* task that is below. Detailed descriptions of both the representation and the learning techniques can be found in [25].

Handling of flexible material Learning both the meaning of symbols and actual actions from instructions provided by a knowledgeable agent (possibly a human teacher) is an important capability especially for robot systems. In our application, the task is to have three robots (Puma 260s) cooperatively handle a flexible bag. Finally, the robots will have to pick up the bag from a conveyor belt and put it into a box. Due to the flexibility of the liquid-filled bag, this is a difficult task which can also not be performed easily by a single robot.

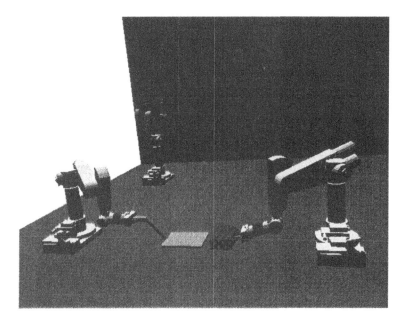

Fig. 6. Robot2 fixes the bag to allow robot1 to lift it.

Both for picking up the bag and for putting it down, two robots have to cooperate. One of the robots is equipped with a shovel that can be used to lift the bag. However, to load and unload the shovel, another robot has to assist, in order to keep the bag in place. Figs. 6–9 show the robots performing this task.

5 Summary and Conclusion

Throughout this paper, the individual learning tasks existing within Multi-Agent Systems have been analyzed with respect to their requirements regarding the communication between teacher and instructed agent(s). To support this analysis, a framework based on system theory and state-space models has been employed.

Learning requires communication and depends crucially on a mutual understanding between teacher and instructed agent. Within the developed framework,

1. the need for an ontology that is common to the teacher and the instructed agent, and

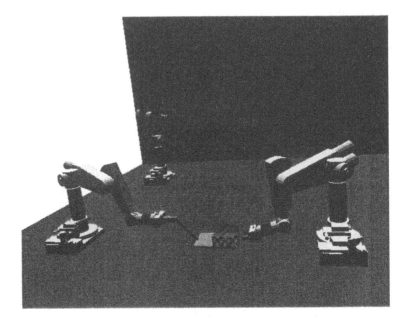

Fig. 7. Robot2 retracts from the bag.

2. the need to extend this ontology in case agents or teams of agents develop new capabilities, in case additional agents join the Multi-Agent System, and for the specification of and negotiation about tasks unknown so far,

have been made explicit. The necessity to explain newly generated symbols to agents has been identified, and a proposal to realize such explanation via "demonstration and observation" has been presented.

As a conclusion, we have seen that communication and the generation of symbols for communication are key components of learning, and, especially, of learning in Multi-Agent Systems. However, while the presented example has shown that agents and agent teams can learn the meaning of symbols by means of explanation, an aspect that has not been treated within this paper is the one of *interpretation* of non-verbal (non-symbolic) messages. This aspect, however, is crucial to fully exploit the potential of learning from demonstration and learning via imitation. Therefore, it will be one focus of our future work.

Acknowledgement

This work has been performed at the Institute for Real-Time Computer Systems & Robotics, Prof. Dr.-Ing. U. Rembold and Prof. Dr.-Ing. R. Dillmann, Department of Computer Science, University of Karlsruhe, Germany. The authors would like to thank the anonymous reviewers for their helpful comments.

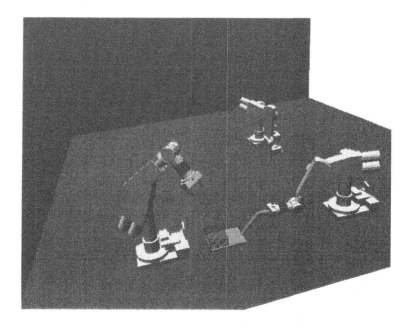

Fig. 8. Robot1 puts the bag down, and robot3 approaches to fix the bag.

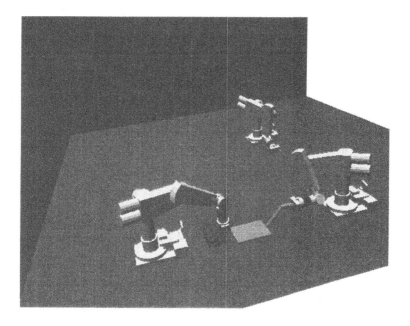

Fig. 9. Robot1 retracts while robot3 fixes the bag. The task is accomplished.

References

1. M. Accame and F.G.B. De Natale. Neural tuned edge extraction in visual sensing. In *Third European Workshop on Learning Robots*, Heraklion, Kreta, 1995.
2. H. Asada and S. Liu. Transfer of human skills to neural net robot controllers. In *IEEE International Conference on Robotics and Automation*, pages 2442 – 2448, Sacramento, 1991.
3. R. D. Beer. A dynamical systems perspective on agent-environment interaction. *Artificial Intelligence*, 72:173–215, 1995.
4. A. I. Cypher. *Watch what I do – Programming by Demonstration*. MIT Press, Cambridge, 1993.
5. N. Delson and H. West. The use of human inconsistency in improving 3D robot trajectories. In *IEEE/RSJ International Conference on Intelligent Robots and Systems*, pages 1248 – 1255, München, 1994.
6. J. Demiris and G. Hayes. Imitative learning mechanisms in robots and humans. In *Fifth European Workshop on Learning Robots*, pages 9 – 16, Bari, 1996.
7. A. Farquhar, R. Fikes, W. Pratt, and J. Rice. Collaborative ontology construction for information integration. Technical report, Stanford University, 1995.
8. Tim Finin, Rich Fritzson, Don McKay, and Robin McEntire. KQML - A language and protocol for knowledge and information exchange. Technical Report CS-94-02, Computer Science Department, University of Maryland and Valley Forge Engineering Center, Unisys Corporation, Computer Science Department, University of Maryland, UMBC Baltimore MD 21228, 1994.
9. M. R. Genesereth, R. E. Fikes, et al. Knowledge interchange format, version 3.0 reference manual. Technical Report Logic-92-1, Computer Science Department, Stanford University, 1992.
10. J. H. Gennari, P. Langley, and D. Fisher. Models of incremental concept formation. *Artificial Intelligence*, 40:11 – 61, 1989.
11. Stevan Harnad. The symbol grounding problem. *Physica D*, 42:335–346, 1990.
12. G. Hayes and J. Demiris. A robot controller using learning by imitation. In *International Symposium on Intelligent Robotic Systems*, pages 198 – 204, Grenoble, 1994.
13. S. Jain and A. Sharma. Team learning of formal languages. In *Agents that learn from other agents: Proceedings of the ICML'95 Workshop*, Tahoe City, California, 1995.
14. M. Kaiser, M. Deck, and R. Dillmann. Acquisition of basic mobility skills from human demonstrations. *Studies in Informatics and Control*, 5(3):223 – 234, September 1996.
15. M. Kaiser and R. Dillmann. Building elementary robot skills from human demonstration. In *IEEE International Conference on Robotics and Automation*, Minneapolis, Minnesota, USA, 1996.
16. M. Kaiser, A. Retey, and R. Dillmann. Designing neural networks for adaptive control. In *IEEE International Conference on Decision and Control*, pages 1833 – 1839, New Orleans, 1995.
17. M. Kaiser and M. Riepp. Learning from undiscounted delayed rewards. In *9. Fachgruppentreffen der GI Fachgruppe 1.1.3 Maschinelles Lernen*, pages 65 – 70, Chemnitz, 1996.
18. V. Klingspor, K. Morik, and A. Rieger. Learning concepts from sensor data of a mobile robot. *Machine Learning*, 23(2/3):305 – 332, 1996.

19. Th. Längle, T. C. Lüth, and U. Rembold. A distributed control architecture for autonomous robot systems. In H. Bunke, H. Noltemeier, and T. Kanade, editors, *Modelling and Planning for Sensor Based Intelligent Robot Systems*. World Scientific, 1995.

20. L. J. Lin. *Reinforcement learning for robots using neural networks*. PhD thesis, Carnegie Mellon University, School of Computer Science, 1993.

21. C. Malcolm and T. Smithers. Symbol grounding via a hybrid architecture in an autonomous assembly system. *Robotics and Autonomous Systems Special Issue on Designing Autonomous Agents*, 6(1,2), 1990.

22. M. J. Mataric. Issues and approaches in the design of collective autonomous agents. *Robotics and Autonomous Systems*, 16(2):321 – 331, 1995.

23. D. A. Pomerleau. Efficient training of artificial neural networks for autonomous navigation. *Neural Computation*, 15:88 – 97, 1991.

24. P. Reignier, V. Hansen, and J.L. Crowley. Incremental supervised learning for mobile robot reactive control. In *Intelligent Autonomous Systems 4*, pages 287 – 294, Karlsruhe, 1995.

25. O. Rogalla. Lernen in Multi-Agenten-Systemen. Master's thesis, Universität Karlsruhe, Fakultät für Informatik, Institut für Prozeßrechentechnik und Robotik, 1996.

26. J. Schmidhuber. A general method for multi-agent reinforcement learning in unrestricted environments. In S. Sen, editor, *AAAI Spring Symposium on Adaptation, Coevolution and Learning in Multiagent Systems*. AAAI Press, 1996.

27. S. Sen, editor. *AAAI Spring Symposium on Adaptation, Coevolution and Learning in Multiagent Systems*. AAAI Press, 1996.

28. K. T. Simsarian and M. J. Mataric. Learning to cooperate using two six-legged mobile robots. In M. Kaiser, editor, *Proceedings of the 3rd European Workshop on Learning Robots (EWLR-3)*, Heraklion, Crete, Greece, April 1995.

29. R. G. Smith and R. Davis. Frameworks for cooperation in distributed problem solving. *IEEE Transactions on Systems, Man, and Cybernetics*, 11(1):61–70, 1981.

30. A. Ude. Trajectory generation from noisy positions of object features for teaching robot paths. *Robotics and Autonomous Systems*, 11:113 – 127, 1993.

31. G. Weiss and S. Sen, editors. *Adaptation and Learning in Multi-Agent Systems*. Springer-Verlag Berlin, Heidelberg, New York, 1995.

Investigating the Effects of Explicit Epistemology on a Distributed Learning System

Nicholas Lacey (njl95@aber.ac.uk)[1], Keiichi Nakata[2] and Mark Lee[1]

[1] University of Wales, Aberystwyth, UK
[2] Oxford Brookes University, Oxford, UK

Abstract. Epistemology is concerned with obtaining a formal, logically coherent definition of knowledge. This paper is concerned with the theoretical and practical aspects of designing a DAI system in which the epistemological foundations of each agent are explicitly stated. The agents learn by modifying the definitions in their ontology in the light of the data they receive from the external world. We intend to show that agents using different epistemological theories will acquire different beliefs, despite being given exactly the same information. This implies that the epistemological foundations of each agent should be taken into account when designing a DAI system.

1 Introduction

Epistemology is the area of philosophy which is concerned with studying theories of truth, knowledge, and belief. Throughout the history of philosophy, epistemologists have developed many intricate theories concerning the nature of truth and knowledge. The differences produced by adopting one theory or another can be far-reaching. What would count as knowledge under one theory, will not necessarily count as knowledge under a different theory.

While these theories and their implications have been well studied by epistemologists, their practical applications have yet to make their way into Distributed Artificial Intelligence (DAI). The purpose of this research is to outline a methodology for designing a distributed computer system which is *explicitly* based on particular epistemological theories, and which allows these theories and their effects to be compared. This will allow the epistemological foundations of individual agents to be matched to the tasks they will be carrying out.

2 Epistemological Background

The purpose of this section is to outline very briefly the issues on which this research is based. Epistemology is concerned with producing a formal account of the concept of knowledge.

The generally accepted definition of knowledge [8] is known as the *tripartite definition*, which is given as follows:

Definition 1 : Knowledge *Agent* A *knows proposition* p *if and only if (iff):*

1. p *is true*
2. A *believes* p
3. A*'s belief that* p *is justified*

Thus, the factors that determine whether or not a particular belief can be classed as knowledge are the *truth* and *justification* of the belief.

The theory of truth which comes the closest to capturing our common sense intuitions concerning the nature of truth is known as the *correspondence* theory, as supported by Bradley and Swartz [6]. They define the correspondence theory of truth as follows:

Definition 2a : Truth - Correspondence *A proposition* p *is true iff the actual state of the world is as* p *asserts it to be.*

According to this theory, "Grass is green" is true because it describes an actual state of affairs, i.e. grass really is green. The fact that the correspondence theory defines the truth of a proposition in terms of whether or not it describes a state of affairs which obtains in the real world has led to the correspondence theory sometimes being referred to as the *realist* theory of truth. Realist theories [5] concerning entity *x* hold that "there really are *x*'s ", even if the existence of *x* makes no difference to us. Aune [2] describes this difference in terms of the difference between theories which hold reality to be fundamentally different from our perceptions of it, and those that do not. Thus, a realist theory of truth would hold that a statement could *really* be true, even if the truth of this statement is completely unknown to us. An anti-realist would deny this claim, as they would reject evidence-transcendent propositions as meaningless. An anti-realist would not necessarily reject the claim that it is possible for an evidence-transcendent external world to exist, but they would reject the realist position that it could be possible to hold meaningful beliefs about it.

The main problem with the correspondence, or realist, theory of truth is that in order to know that a proposition is true, we have to be able to know that the proposition is true in the real world. Given that Descartes [11] has shown us that our limited sensing capabilities are prone to errors, we cannot claim absolute knowledge concerning the state of the real world.

An alternative to defining truth in terms of correspondence is to define it in terms of *coherence*. The coherence theory of truth defines truth as follows [8]:

Definition 2b : Truth - Coherence *A proposition is true iff it is a member of a coherent set.*

Just as the correspondence theory of truth could be described as a realist theory, so the coherence theory of truth can be described as an anti-realist theory. This is because the truth of a proposition is not dependent upon the actual state of the world. Thus, the proposition "Grass is green" is true because my belief that grass is green coheres with my view of the world, rather than because it describes an actual state of affairs.

Correspondence theorists have objected to the coherence theory of truth on the grounds that two or more mutually contradictory sets could be equally internally coherent, which would result in multiple versions of "the truth". This is known as the *"plurality objection"*, and while it seems at first sight to be an effective criticism of coherence theories, there is in fact very little substance to it. This is because, as the sets are mutually inconsistent, they would not be believed at the same time by the same agent, and so the coherence theorist does not have to accept the absurd proposition that two contradictory propositions can both be true at the same time. The plurality objection is only effective for someone who accepts that "real" evidence-transcendent truth exists. Anti-realists deny this claim.

Justification is the process whereby a belief is supported by other beliefs. Exactly how this process works is the subject of debate. As Dancy [8] describes, there are two major approaches to justification, namely the foundationalist approach and the coherence approach. The foundationalist theory of justification holds that our beliefs are organised in a structure that is comparable to a pyramid, whereby we have a set of basic beliefs on which our non-basic beliefs are based. Thus, this theory would define justification as :

Definition 3a : Justification - Foundationalism *A non-basic belief is justified iff it is entirely based on basic beliefs. Basic beliefs require no justification from other beliefs.*

The main problem with foundationalism is that it requires that there are such things as basic beliefs which are either self justifying or require no justification from other beliefs. Given the fallibility of our senses, the existence of self-evident basic beliefs seems at best doubtful. There are different flavours of foundationalism which attempt to overcome this problem in slightly different ways. All foundationalist theories, however, hold that the justificational relationship between basic and non-basic beliefs is asymmetrical. Thus, a basic belief may justify a non-basic belief, but a non-basic belief may never, even in part, justify a basic belief.

In contrast to the foundationalist theory of justification, the coherence theory of justification holds that justification is based on coherence with other beliefs rather than foundation on basic beliefs. Just as the foundationalist approach could be represented as a pyramid, so the coherence approach to justification can be represented as a raft, whereby our beliefs mutually support each other, and no single belief is supported by anything other than other beliefs.

Dancy describes the coherence theory of justification as follows :

Definition 3b : Justification - Coherence *A belief is justified to the extent to which the belief-set of which it is a member is coherent.*

The fundamental difference between coherence and foundationalist theories of justification is that, according to coherence, *any* belief may play a part in the justification of *any other* belief. Thus, the justificational asymmetry which is essential to foundationalism is not present in coherence.

As with foundationalism, there are different strengths of coherence theory. An important consequence of strong coherence, especially when combined with an anti-realist [5] metaphysical position, is that *no* beliefs are immune from revision. This conclusion follows from Quine's rejection of the analytic/synthetic distinction [15].

Thus, any account of knowledge which accepts the tripartite definition will necessarily also have to provide an account of truth and justification, while also refuting sceptical arguments. This is an extremely complex task, which it would not be relevant to attempt here [3]. The important point to note is that the philosophical concept of knowledge is a complex structure encompassing truth, belief, and justification. This means that, even among epistemologists who accept the tripartite definition, there are radically different accounts of what knowledge actually is. It follows that which account of knowledge we accept does make a difference as to what counts as knowledge, and how the beliefs which we accept as knowledge should be revised in the light of new information.

3   Making Epistemology Explicit in AI

The principal way in which the AI use of the word "knowledge" differs from the philosophical use of the term concerns the concept of truth. In [12] (page 45), Levesque and Brachman write :

> *Notice that typical of how the term "knowledge" is used in AI, there is no requirement of truth. A system may be mistaken about the colour of the sky but still be knowledge-based. "Belief" would perhaps be a more appropriate term, although we follow the standard AI usage in this paper.*

The approach described by Levesque and Brachman reflects the implicit assumption that truth is something that is defined in terms of correspondence to an external reality, and hence is beyond the scope of a computer system, i.e. a correspondence theory of truth is being implicitly assumed here.

The aim of our research is to examine the effects of using different theories of truth and justification. In particular, the adoption of an anti-realist metaphysical position combined with a coherence-based theory of truth and justification would have a dramatic effect on the methods used by a system to acquire knowledge, the architecture required for a system to process knowledge, and the results that would be achieved by the system. The understanding of the implications of such different theories will allow optimal matches to be made between epistemological foundations and problem domains.

In order to accomplish this objective, a system which allows the effects of using a particular set of epistemological foundations has been designed. Techniques which allow the computational modelling of justification have already been developed, such as Doyle's Truth Maintenance Systems [9] and work in the field of belief revision [10]. For this reason, we will concentrate here on techniques which will allow the concept of truth to be made explicit.

---

[3] For more information, see [8]

4 Description of System

The purpose of this section is to describe briefly the system that has been designed to address the issues outlined above, with particular emphasis placed on the parts of the system which deal with learning.

The system is based around a multiple agent architecture, as shown in figure 1. There is a strong distinction between the *external world* and the *internal world*. For a particular agent, the external world is everything which is external to that agent.

The system is being implemented using SICStus Prolog [7]. SICStus Prolog contains library functions supporting object-oriented programming, as well as inter-process communication via Linda [1]. This allows the agents to run as separate Prolog processes, communicating via a shared tuple space to which all the agents can read and write.

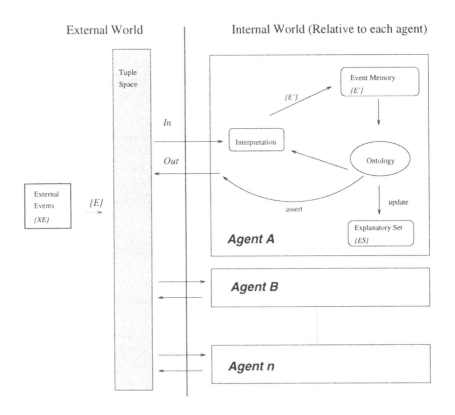

Fig. 1. System Architecture

4.1   The External World

The external world contains the domain that is to be modelled. Information concerning external events is placed onto an area of shared memory called the "tuple space", from where it can be read by the agents. Events occur in the external world in discrete ordered time-slices. Thus, at any given time t_n, a set of events $\{XEt_n\}$ will be occurring in the external world. A subset of $\{XEt_n\}$, $\{Et_n\}$, is entered on to the tuple space. Events are entered onto the tuple space using a syntax based on situation theory, as described in [3]. The format used is:

```
[external-data-source,[state,object,i]]
```

where `external-data-source` represents the source of the information, which will either be an agent or input from sensors, and where `i` is a truth indicator which can be set to `1`, representing `true`, or `0`, representing `false`.

For example, if at a particular time `sensor-1` indicates that object `car` is in state `stopped` and object `traffic-light` is in state `red`, this would be entered onto the tuple space in the following format :

```
[

  [sensor-1,[stopped,car,1]]
  [sensor-1,[red,traffic-light,1]]
]
```

Now, both of these events may entail other events, such as `[moving,car,0]` and `[green,traffic-light,0]`. However, as the information entered onto the tuple space is a subset of the events occurring in the external world, it will be up to the agents to decide what else is going on in the external world. This is carried out based on the rules stored in their ontology, as described in section 4.2.

4.2   The Internal World

The internal world of each agent is private to that agent. Each agent in the system has its own independent ontology. An ontology can be thought of as a hierarchical formalisation of a conceptualisation, within which all the assumptions that were made during the design of the KBS have been made explicit. As described in [13], ontologies are constructed on several levels. The higher levels contain abstract definitions which are used in the lower levels to record current run-time assertions.

Raw data describing a set of external events occurring at a particular time $\{Et_n\}$ is placed in the tuple space. This data is then received by the individual agents, who interpret the data according to their current ontology. The interpreted data $\{E't_n\}$ is then stored in the agent's memory of events $\{E'\}$. All agents receive identical input from the external world, although they may choose to interpret it differently.

The concepts within the agents' ontology are defined in terms of the *logical sufficiency* (λ) and *logical necessity* $(\overline{\lambda})$ of the properties that must be present in order for the concept to obtain. The PROSPECTOR system [17] uses a similar system of λ and $\overline{\lambda}$ values for describing rule strengths. This allows the representation of complex asymmetrical relationships between concepts. The scale for λ and $\overline{\lambda}$ values used here is based on the *odds-likelyhood* (1) and *odds-unlikelyhood* (2) formulations of Bayes rule. [14]

$$O(H \mid E) = \lambda O(H) \tag{1}$$

$$O(H \mid \overline{E}) = \overline{\lambda} O(H) \tag{2}$$

where $O(H)$ represents the prior odds for hypothesis H, $O(H \mid E)$ the revised odds for H given evidence E, and $(H \mid \overline{E})$ the revised odds for H given the absence of evidence E. These scales reflect the extent to which support for a hypothesis is increased or decreased given the presence or absence of evidence. The arbitrary scale used here uses 100 for maximal sufficiency, 1 for neutral sufficiency or necessity, and 0 for maximal necessity. When the prior odds are multiplied by the λ or $\overline{\lambda}$ value, the revised odds will increase or decrease accordingly.

The relationships between concepts in the ontology are stored using rules which supply the λ and $\overline{\lambda}$ values relating to the *IF* and *THEN* part of that particular rule. For the lower-level rules, these values are calculated not only for the current time slice t_n, but also for an arbitrary number of time slices into the future. The number of slices by which rules look ahead is determined by the size of the *Rule Window*, W_R. Hence, each rule maintains λ and $\overline{\lambda}$ values that describe the relationship between the rule's *IF* and *THEN* parts between t_n, and t_{n+W_R}. Figure 2 shows the structure and component parts of a rule.

For example, the rule shown in table 1 represents the fact that if object `car` is in state `moving` at time t_n, this does not increase support for the hypothesis that object `traffic-light` will be in state `red` at time t_n or t_{n+1}, but this condition is maximally sufficient and necessary support for the hypothesis to apply at time t_{n+2}:

IF	THEN	$\lambda,\overline{\lambda}$ values		
		t_n	t_{n+1}	t_{n+2}
`[[moving,car,1]]`	`[red,light,1]`	`[1,1]`	`[1,1]`	`[100,0]`

Table 1. An example rule

The temporal nature of these rules makes them very flexible, and allows inductive learning techniques to be used to acquire temporal information, as described in section 5. However, while this format is useful for low-level rules, it is not appropriate for high-level meta-rules, such as those concerning truth and knowledge. Therefore, while meta-rules use λ and $\overline{\lambda}$ values, they do not maintain these values over multiple time slices. Instead, these values refer to the current time slice (t_n) only.

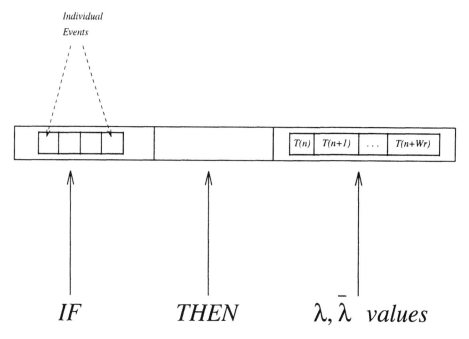

Fig. 2. The structure of a rule

The use of high level meta-rules makes it possible to give an agent a particular set of epistemological assumptions by placing meta-rules for concepts such as coherence, correspondence, truth, justification, and knowledge, at a high level in its ontology. For example, two agents could be given different theories of truth by using different definitions for truth. Thus, an agent could be given a concept of truth whereby correspondence with an external reality were both necessary and sufficient for truth to obtain. This would be defined as follows :

Truth
[corresponds,external-reality,1]
λ = 100 $\bar{\lambda}$ = 0

An agent with this definition of truth would try to make its beliefs correspond with data it received by altering its beliefs as required. The concepts of corresponds and external-reality would both be defined elsewhere in the ontology using definitions based on Definition 2a.

While an agent using a correspondence-based view of truth will alter its beliefs to fit in with the data it receives, an agent with a coherence-based view of truth would be more concerned about maximising the coherence of its ontology than about accepting new information, and so would reject information which did not increase the coherence of its ontology. An agent could be given a coherence-based view of truth using a definition such as :

Truth
[coheres,current-ontology,1]
$\lambda = 100 \ \overline{\lambda} = 0$

The concept of coherence would also be defined in the agent's ontology, using a definition based on Definition 2b. In general, an agent will alter its low-level definitions in order to preserve its high-level definitions. However, this is not to say that it would be impossible for a coherence-based agent to change its beliefs. Indeed, a coherence-based agent would readily change its beliefs if doing so lead to a more coherent ontology. If this agent were also using a coherence-based theory of justification, as would be expected, then *any* of its beliefs would be revisable, including its high-level epistemological definitions. This means that if an agent is faced with a choice between revising a high-level belief and revising many low-level beliefs, it might chose to revise the high-level belief if doing so increases the overall coherence of the ontology.

For example, if someone walks into my office and tells me that some Martians have landed in the car-park outside the building, I will probably choose not to believe them. Instead, I will explain the evidence that is available to me by concluding that the person was lying, for whatever reason. If many more people, all of whom have been reliable in the past, also come and tell me that some Martians have landed in the car-park, and I also hear about this on the radio and television, then I am faced with a choice. Firstly, I can maintain my current belief that Martians do not land in car-parks. However, in order to do this I would have to conclude that sources of information that had proved reliable in the past had suddenly ceased to be reliable. This course of action would not necessarily increase the overall coherence of my belief-system, as all the information I had received from these people in the past would now have to be re-evaluated. Alternatively, the overall coherence of my belief system might be increased by maintaining my belief that these people do, in general, tell the truth, and by accepting that something out of the ordinary has indeed occurred in the car-park.

Thus, a coherence-based agent is perfectly capable of revising its beliefs, even high level ones, as long as doing so increases the coherence of its ontology. Thus, if such an agent decided that the coherence of its ontology would be increased by altering its definition of what "truth" is, then this is what it would do. This approach to belief revision is similar in some ways to the "coherence" model suggested by Gärdenfors [10], except that belief revision is concerned with justification, as opposed to the nature of truth itself.

5 Learning

The purpose of our research is to investigate the extent to which using different epistemological theories affects the manner in which agents form rules concerning the operation of the external world. It is therefore essential that the agents used in this system are able to form their own rules based on the information that is available to them.

Information concerning which concepts in the ontology are related to which other concepts, and the strengths of these relationships, are stored using rules described in section 4.2. Any information learned by the agents must be transformed into this format. This section describes the techniques used by the system to allow the agents to induce their own rules using the data available to them, and then to transform the induced rules into a format which can be used by their ontology.

5.1 Inductive Learning

The agents need to be able to learn the extent to which concepts in their ontology are necessary and sufficient for other concepts to apply. In order to do this, we have used the ID3 inductive learning algorithm, as described by Quinlan in [16][4]. Given a set of training data containing sets of attribute-value pairs that have been classified as constituting a positive or negative example of a particular event, ID3 constructs a decision tree which can be used to classify new data. In the current system, the size of the training set available to ID3 is determined by the size of the *Learning Window* (W_L). In order to induce rules concerning when a particular `object` is in a particular `state`, all the time slices between the current time t_n and t_{n-W_L} are labelled as constituting a positive example (if the event $[state, object, 1]$ occurs at that time) or a negative example (if it does not). ID3 then constructs a decision tree which classifies the data in the training set into positive or negative examples depending on the attributes of the data.

The fact that data in the external world being assumed here occurs during discrete temporal intervals makes the process of creating the training data very straightforward. However, the data in the external world is *ordered*, whereas ID3 does not use the order in which data is presented to it, as it has no concept of example data sets preceding or following one another. Hence, important temporal data is lost in the process of preparing the data for ID3.

The solution we have adopted to this problem was to place temporal information into *every* low-level rule, as described in section 4.2. This means that all ID3 has to do is to produce *IF* and *THEN* parts to rules, based on the data it perceives as being unordered. These *IF* and *THEN* pairs are then placed into the rule template shown in figure 2, and the λ and $\overline{\lambda}$ values for all time-slices between the current slice t_n and the end of the rule window t_{n+W_R} are then automatically calculated.

5.2 The Learning Algorithm

Once the agents have interpreted the information received from the tuple space, it is stored in memory. The number of past time frames for which the system records events is determined by the size of the *Memory Window* (W_M) variable. Figure 3 shows how events are passed into ID3. The letters in the *Data* column represent events, in $[state, object, i]$ format, such as $[moving, car, 1]$.

---

[4] The Prolog implementation of ID3 used in this system was written by Luis Torgo.

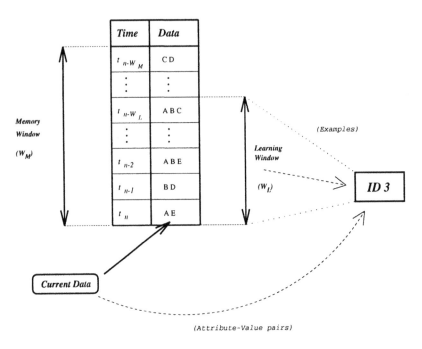

Fig. 3. Passing data from the Learning Window to ID3

Newly interpreted data arrives at time t_n and is stored in memory. The subset of past events which is used to derive the examples to be passed to ID3 is determined by the size of W_L.

All the events occurring at t_n are then passed in to ID3 as the attribute-value pairs for which it is to induce rules one at a time. It is expected that many of the rules produced at this stage will be duplicates of rules that are already stored in the agent's ontology. While it would be possible to prevent attribute-value pairs for which rules already exist from being passed to ID3, this would prevent the agents from producing new, possibly more accurate, rules for events which already have rules linking them. It was decided that the increased diversity of rules produced by this approach was worth any decrease in efficiency caused by producing duplicate rules.

Thus, the manner in which the various agents will induce new rules is identical, in that they will use use ID3 to induce rules based on their memories of past events. The newly induced rules will then have temporal information added to them. The differences between the rules learned by the agents will be due to the fact that they will not necessarily have the same events stored in memory, despite the fact that they will all be given exactly the same information from the tuple space.

6   Experiments and Expected Results

This section describes the experiments that will be carried out using the system, and the results we expect to obtain. It will be split into two sections. Section 6.1 dealing with learning within an individual agent, and section 6.2 dealing with the interaction between agents.

6.1   Learning from Observation

In order to test this stage of the learning process, agents will be given data concerning a car and a traffic light, both of which will be in different states at different times. The agents will be expected to induce rules linking the states of the traffic light to the motion of the car. While this is an extremely simple example, it is sufficient to show how data is transferred from memory into ID3.

Table 2 shows the state that an agent's memory might be in at t_4.

Time	Events
t_1	[moving,car,1] [red,light,0] [green,light,1]
t_2	[moving,car,0] [red,light,1] [green,light,0]
t_3	[moving,car,1] [red,light,0] [green,light,1]
t_4	[moving,car,0] [red,light,1] [green,light,0]

Table 2. The format of the example data in agent memory

All the events occurring at the current time t_4 are passed into ID3 one at a time. In this example, we will go through the process of inducing rules to decide when [moving,car,0] applies.

The parameters used in process of preparing the examples used by ID3 are $W_M = 4$, $W_L = 4$, and $W_R = 3$. A time slice will be labelled as a positive example if it contains the event [moving,car,0]. After they have been used to classify the data into positive and negative examples, events concerning [moving,car,i] are removed from the data before it is passed to ID3. Table 3 shows how the data shown in table 2 will be shown to ID3.

When ID3 is called using the data shown in table 3, it produces the following two rules :

[[light,red],0,[leaf(negative)]]

and

Example Number	Class	Example Data
1	Negative	`[[light,red]=0,[light,green]=1]`
2	Positive	`[[light,red]=1,[light,green]=0]`
3	Negative	`[[light,red]=0,[light,green]=1]`
4	Positive	`[[light,red]=1,[light,green]=0]`

Table 3. The format of the ID3 example data

$$[[\text{light},\text{red}],1,[\text{leaf}(\text{positive})]]$$

When these two rules are transformed into the rule template described in section 4.2 and the positive and negative leaf nodes are replaced by the events they represent, we obtain the rules shown in table 4.

IF	THEN	$\lambda,\overline{\lambda}$ values		
		t_n	t_{n+1}	t_{n+2}
`[[red,light,0]]`	`[moving,car,1]`	`[100,0]`	`[1,1]`	`[100,0]`
`[[red,light,1]]`	`[moving,car,0]`	`[100,0]`	`[1,1]`	`[100,0]`

Table 4. The new rules with temporal information added

The first rule represents the fact that if the light is not red at t_n, then the *a priori* odds that the car is moving at t_n and t_{n+2} should be multiplied by 100. On the other hand, if the light is red at t_n then the *a priori* odds that the car is moving at t_n and t_{n+2} should be multiplied by 0. The odds for t_{n+1} are not affected by this rule. The second rule provides a corresponding relation between the light being red and the car not moving.

While this is a very simple example, it does show how the system uses data from an agent's memory to induce rules using ID3, to which temporal information is then added. Rules induced using more complex data might involve more than one condition in the *IF* part of the rule, and the λ and $\overline{\lambda}$ values other than the maximal 100 and 0.

6.2 Agent Interaction and Learning

This section provides an example of the manner in which we expect the agents to learn by interacting with each other. It is expected that agents holding different epistemological beliefs will form different beliefs concerning the operation of the external world, even though they have received exactly the same information. A system of slots similar to those described in [4] within each agent's self-model will allow the agents to keep track of the beliefs of the other agents. This disagreement between agents will lead to inter-agent communication which will form an additional source of information for the agents.

The system will be used to model the operation of a manufacturing process in which the temperature of a furnace is controlled by opening or closing the valve which supplies fuel to the furnace. Sensors record the temperature of the furnace and the state of the valve, and pass their output to the agents, via the tuple space. In this example, the sensor input represents the external world.

The agents are given identical definitions of all the concepts in their ontologies, except that Agent A has a correspondence-based view of truth whereas Agent B has a coherence-based view of truth. New information will then be given to them, which will force them to revise their beliefs to maintain consistency. The agents will have to organise their ontologies so as to best explain the evidence that is available to them. The methods used by the agents to re-organise their beliefs will depend upon the epistemological assumptions they are using.

Table 5 shows the result of an analysis of the interaction between these two agents as new data is entered into the tuple space by the sensors.

Time	Agent A (Correspondence)	Agent B (Coherence)	Tuple Space
t_1			`[sensor-1,[closed,valve,1]]`
t_2	If valve is closed temp must be low. Post this on tuple space	If valve is closed temp must be low. Post this on the tuple space.	`[agent-A,[low,temp,1]]` `[agent-B,[low,temp,1]]`
t_3	Notes that Agent-B is reliable.	Notes that Agent-A is reliable.	
t_4			`[sensor-1,[open,valve,1]]` `[sensor-1,[low,temp,1]]`
t_5	Revises its beliefs so that the temperature can be low while the valve is open.	The new data is inconsistent. This means the sensor is unreliable.	`[agent-A,[open,valve,1]]` `[agent-A,[low,temp,1]]` `[agent-B,[reliable,sensor-1,0]]`
t_6	The input from Agent-B is disregarded. Notes that Agent-B is unreliable.	Notes that Agent-A is unreliable.	

Table 5. Example interaction between two Agents

The first entry onto the tuple space states that the valve is **closed**. Both agents agree that if the valve is `closed`, then the temperature of the furnace must be `low`. Both agents send this assertion onto the tuple space. Both agents then receive this information from each other, and both decide that the other agent is a reliable source of information.

At t_4, the sensors indicate that the temperature of the furnace is `low`, while the valve is `open`. This information is inconsistent with what the agents already believe. What is interesting here is the manner in which the agents react to this information.

Agent A will attempt to re-organise its beliefs so as to explain the new information, and so forms new definitions of `open` and `low` such that the valve can be `open` while the temperature is `low`. It also sends the fact that it has done this on to the tuple space. Agent B, on the other hand, will reject this information as inconsistent, and interprets this information by deciding that the input is unreliable. This conclusion is also sent to the tuple space.

Both agents now know how the other agent reacted to the inconsistent information. Agent B's conclusion that the input was unreliable is not acceptable to Agent A, as Agent A determines the truth of its beliefs in relation to how well they correspond with the readings produced by the sensors. Agent A therefore decides that Agent B is unreliable. Any future information placed on the tuple space by Agent B will be treated with scepticism by Agent A. Agent B, on the other hand, cannot accept Agent A's conclusion that its ontology was incorrect, as Agent B measures the truth of beliefs in accordance with how well they cohere with what it already believes. Agent B therefore decides that agent A is unreliable.

In a control system such as the one used in this example, agents using different epistemological foundations present different explanations to explain the evidence from the sensors. In other words, they are forming different hypotheses to explain the operation of the external world. What is *really* going on in the external world is unknown to the agents, just as it is to us. Which set of epistemological assumptions an agent uses determines the manner in which it will form hypotheses in the light of new data. By using a Multi-Agent System in which agents are based on different epistemological foundations, we are increasing the chances that at least one agent will produce hypotheses which successfully model the problem domain.

This means that, as we are more interested in the *differences* between the agents' beliefs than we are in the similarities between them, we are not expecting that the agents will necessarily come to any agreement as to the operation of the external world. Instead, it is expected that each agent will form its own internally consistent system, will interpret new data in accordance with that system, and will conclude that the agents which do not agree with its interpretations are mistaken. Thus, we are expecting the agents to exhibit local coherence rather than global coherence [4].

In the above example, all the agents are being given exactly the same information, and are playing homogeneous roles in the system. By using agents with different epistemological foundations, we are increasing the diversity of hypotheses produced. However, it would also be possible to place agents with different epistemological foundations into heterogeneous roles in the system, by tailoring an agent's epistemological foundations to the particular task that the agent will be carrying out. It is expected that agents who need to react quickly to a reliable data source would benefit from a correspondence-based approach, while agents who will be trying to form general rules or are working with unreliable data would perform better using a coherence-based approach.

Thus, it is expected that agents based on different epistemological theories will acquire different beliefs to explain the same information. This means that using a Multi-Agent System in which agents are using different epistemological approaches will increase the diversity of solutions found, and increase the chances of an accurate model being formed. It is also expected that certain epistemological foundations will be better suited to certain types of learning problem than others.

7   Conclusions

The main conclusion that is expected to be drawn from this research is that agents using different epistemological theories will acquire different beliefs despite being presented with *exactly* the same information. This will show that epistemological assumptions do affect the learning and knowledge-processing behaviour of agents, and as such should be *explicitly* taken into account when a Multi-Agent System is being designed so that the optimal set of epistemological foundations can be matched to the tasks of individual agents.

References

1. Jonas Almgren, Stefan Andersson, Mats Carlsson, Lena Flood, Seif Haridi, Claes Frisk, Hans Nilsson, and Jan Sundberg. *SICStus Prolog Library Manual*. Swedish Institute of Computer Science, Kista, Sweden, January 1993.
2. Brune Aune. *Metaphysics : The Elements*. Basil Blackwell, 1985.
3. Jon Barwise and John Perry. *Situations and attitudes*. MIT Press, Cambridge, Mass, 1983.
4. N.R. Jennings Benedita Malheiro and Eugénio Oliveira. Belief revision in multi-agent systems. In A. Cohn, editor, *Proceedings of ECAI 94, 11th European Conference on Artificial Intelligence*, pages 294–298. ECAI 94, 11th European Conference on Artificial Intelligence, John Wiley and Sons, Ltd, 1994.
5. Simon Blackburn. *Spreading the Word : Groundings in the Philosophy of Language*. Clarendon Press, Oxford, 1984.
6. Raymond Bradley and Norman Swartz. *Possible worlds : an introduction to logic and its philosophy*. Basil Blackwell, Oxford, UK, 1979.
7. Mats Carlsson and Johan Widén. *SICStus Prolog User's Manual*. Swedish Institute of Computer Science, Kista, Sweden, January 1993.
8. Jonathan Dancy. *Introduction to Contemporary Epistemology*. Basil Blackwell, 1985.
9. John Doyle. A truth maintenance system. *Artificial Intelligence*, 12:231–272, 1979.

10. Peter Gärdenfors. *Belief Revision : An Introduction*, volume 29 of *Cambridge Tracts in Theoretical Computer Science*. Cambridge University Press, Cambridge, 1992.

11. E.S. Haldane and G.R.T. Ross. *The Philosophical Works of Descartes*. Cambridge University Press, Cambridge, 1955.

12. Hector J. Levesque and Ronald J. Brachman. A fundamental tradeoff in knowledge representation and reasoning (revised version). In *Readings in Knowledge Representation*, pages 41–70. Morgan Kaufmann, Los Altos, California, 1985.

13. Robert Nesches, Richard Fikes, Tim Finin, Thomas Gruber, Ramesh Patil, Ted Senator, and William R. Swartout. Enabling technology for knowledge sharing. *AI Magazine*, Fall, 1991.

14. Judea Peal. *Probabilistic Reasoning in Intelligent Systems : Networks of Plausible Inference*. Morgan Kaufmann, San Mateo, CA, 1988.

15. Willard Van Orman Quine. Two dogmas of empiricism. In *From a Logical Point of View : 9 logico-philosophical essays*, chapter 2, pages 20–46. Harvard University Press, Cambridge, Mass, 1980.

16. J. Ross Quinlan. Learning efficient classification procedures and their application to chess end games. In Ryszard S. Michalski, Jaime G. Carbonell, and Tom M. Mitchell, editors, *Machine Learning : An Artificial Intelligence Approach*, volume 1-2, pages 463–482. Tioga Pub. Co, San Mateo, Calif., 1983.

17. David J. Spiegelhalter. A statistical view of uncertainty in expert systems. In Judea Pearl Glenn Shafer, editor, *Readings in Uncertain Reasoning*, pages 313–331. Morgan Kaufmann, San Mateo, California, 1990.

Subject Index

A

action 96, 98
addressee learning 245
agent 13, 280
agent model 260
alien references 188
anytime algorithm 63
anticipatory agent 62f

B

belief revision 154
Bayesian conditioning 157

C

case-based learning 5
case-based reasoning 6, 180f, 245
co-evolution 87
cognizant failures 69
collective behavior 30
collective case-based reasoning 193, 195
combinatorial explosion 26, 27
communication costs 173
communication in isolated learning 263
communication in teaching 265
communication for learning 266
competing agents 66
contract net learning 204f
contract net protocol 170, 204f, 244
control programs 40
cooperative agents 67
cooperative information systems 11
coordination 118f
credit assignment 214
credit assignment problem 2

D

data mining 16
Dempster's rule of combination 156

detection function 98
distributed case-based reasoning 192, 194

E

effort allocation 104
epistemology 276
evolutionary learning 4
explanation 266

F

federated peer learning 181
focused addressing 245
foreign method evaluation 190

G

game theory 118f
global knowledge 112

H

heterogeneous robots 43
homogeneous robots 43

I

imitation learning 269
individual learning 170
inductive learning 222f, 285
inferential theory of learning 233
interface agents 202, 203, 218

J

justification 278

K

knowledge adaptation 99
KQML 223f, 236

L

learning 110, 112
learning by discovery 205
learning how to learn 84
learning of reliability 154
LEMMING 244
linear regression 143
local knowledge 112

M

manager learning function 213f
mediator 15
message interception 249
mobile methods 191
modular Q-learning 27
meta-learning 84
monolithic reinforcement learning 25f
multiagent block pushing problem 74
multiagent object search 94f
multiagent systems 118
multiagent system model 261

N

Noos representation language 184

O

object search 95f
ontology 16, 281
open environment 11–12, 19, 24
open system 12, 19
opponent model 141, 144
organizational learning 5, 171

P

physical grounding hypothesis 7
Plural Noos agent communication language 188
pursuit problem 26, 87

predator and prey *see* pursuit problem

Q

QCON algorithm 75
Q-learning 27, 74f

R

real team solutions 41f
reinforcement learning 82, 205
reliability 154
robot credit assigment problem 40
robotic soccer 137, 139

S

shared learning 146
shared meaning problem 7
simulation 139f, 158
skill acquisition 269
spatial reference frame 219f
structural credit assignment 73
success-story algorithm 83, 86
success-story criterion 83
symbol grounding problem 7
symbol system hypothesis 7

T

task allocation problem 171
temporal credit assignment 73
theories of truth 276f
traffic signal control 119, 126
truth 8, 277

U

user adaptation 202, 203
user models 203

Springer
and the
environment

Lecture Notes in Artificial Intelligence (LNAI)

Lecture Notes in Computer Science